# Knowledge and Evolution

# Knowledge and Evolution

How Theology, Philosophy, and Science
Converge in the Question of Origins

Michael Chaberek

RESOURCE *Publications* • Eugene, Oregon

KNOWLEDGE AND EVOLUTION
How Theology, Philosophy, and Science Converge in the Question of Origins

Copyright © 2021 Michael Chaberek. All rights reserved. Except for brief quotations in critical publications or reviews, no part of this book may be reproduced in any manner without prior written permission from the publisher. Write: Permissions, Wipf and Stock Publishers, 199 W. 8th Ave., Suite 3, Eugene, OR 97401.

Resource Publications
An Imprint of Wipf and Stock Publishers
199 W. 8th Ave., Suite 3
Eugene, OR 97401

www.wipfandstock.com

PAPERBACK ISBN: 978-1-6667-0207-1
HARDCOVER ISBN: 978-1-6667-0208-8
EBOOK ISBN: 978-1-6667-0209-5

08/09/21

Imprimatur: Rev. Pawel Kozacki. Reg. Prov. 033/21

# Contents

*Preface* | ix

**Chapter I: Religion and Science** | 1

    1. Theology, Philosophy, and Science | 3
        a. The Three Levels of Human Knowledge | 3
        b. Knowledge and Truth | 9
        c. The Two Books | 11
        d. Faith and Reason vs. Faith and Science | 13

    2. Naturalism | 17
        a. Naturalism by Expansion | 17
        b. Naturalism by Rejection | 18

    3. Historical Interdependence between Religion, Philosophy, and Science | 20
        a. The Historiosophy of Auguste Comte | 20
        b. Science, Philosophy, and Religion throughout the Ages | 22

    4. The Limits of Natural Explanations | 34
        a. The Limits of Science in Terms of Space | 35
        b. The Limits of Science in Terms of Time | 49
        c. The Limits of Science and Biological Macroevolution | 54

    5. Galileo and Darwin | 58
        a. The Galileo Affair: A Summary | 58
        Galileo and the Bible | 63

    6. The Solution to an Apparent Conflict between Science and the Bible | 70
        a. "How?" vs. "From Where?" | 70
        b. The Bible and the Origin of Geological Formations | 75
        c. The Progress and the Limits of Science | 78

Chapter I—Summary | 84

**Chapter II: Evolution and Natural Knowledge | 85**

   1. Preliminary Definitions | 85
      a. Species | 85
      b. Evolution | 90
      c. Biological Macroevolution | 94

   2. The Paradoxes of Darwinism | 96
      a. Paradoxes of the Ancients | 96
      b. Species Exist and They Do Not Exist | 98
      c. Natural Selection Simultaneously Builds Up and Ruins Species | 101
      d. The Paradox of the Necessarily Growing Divergence between the Discovery and the Theory | 103
      e. The Paradox of Undefined Species | 105
      f. The Paradox of Random Variation and Natural Selection | 110

   3. Evolution and Metaphysics | 114
      a. An Accidental Change Does Not Generate a Substantial Change | 116
      b. Response to the Objection Raised against the Darwinian Paradoxes | 122
      c. The Problem of Sufficient Cause | 123
      d. The Problem of the Reduction of Causes | 126
      e. Transformation of Species vs. Biological Macroevolution | 130

   4. Evolution and Science | 132
      a. The Fossil Record | 133
      b. Homology | 138
      c. The Classical Explanation of Homology | 141
      d. Survival of the Fittest as Tautology | 142
      e. Epigenetics | 143

   5. Intelligent Design and Neo-Darwinism | 145
      a. "Probabilistic Miracles" | 147
      b. The Explanatory Filter and Specified Complexity | 151
      c. A Universal Probability Bound | 156
      d. Where Does the Explanatory Filter Take Us? | 157
      e. An Agent, a Law, and Chance | 159

      f. Irreducible Complexity | 160
      g. Evolutionary Steps | 162
      h. Is Darwinism More Scientific Than Intelligent Design? | 167
      i. The Limits of Methodological Naturalism | 172
      j. Does Intelligent Design Violate Methodological Naturalism? | 175
      k. Methodological and Ontological Naturalism | 179

  Chapter II—Summary | 180

## Chapter III: Evolution and Christianity | 181

  1. The Arena of the Current Debate | 182
      a. The History of Western Thought—from Paganism to Neo-Paganism | 182
      b. The Four Concepts of the Origin of Species | 186
      c. The Inaccuracy of the Term "Creationism" | 192
      d. The Interpretative Problems with Theistic Evolution | 193
      e. The Amount of Creation in Creation | 195
      f. Who Debates Whom in this Debate? | 200

  2. Christianity and Creation | 203
      a. Two Christian Interpretations of Genesis | 203
      b. Three Stages of Creation | 206
      c. The Nature of the Creative Act | 208

  3. Christianity and Theistic Evolution | 216
      a. Theistic Evolution and the Notion of Creation | 216
      b. Augustine and Theistic Evolution | 217
      c. Can Chance Take Part in Creation? | 221
      d. Theistic Evolution and Historical Realism | 223
      e. Theistic Evolution and Christianity—A Summary | 225

  4. Christianity and Intelligent Design | 226
      a. The Design Inference and the Design Argument | 226
      b. Intelligent Design and the "God of the Gaps" | 229
      c. Intelligent Design and Philosophy of Nature | 232
      d. What's the Problem with Philosophy of Nature? | 236
      e. Philosophy of Nature and Biological Macroevolution | 240

  5. Catholicism and Evolution | 243
      a. Has the Church Ever Condemned Darwin? | 243
      b. Faith and Science in the Current Ecclesiastical Debate | 249
      c. Toward the New Science-Faith Synthesis | 253

  d. The Renewal of the Catholic Theology of Creation | 257
  e. The Benefits of the Renewal | 260
 6. Christianity and Culture | 263
  a. Richard Niebuhr's "Christ and Culture" | 263
  b. Can Christianity be Countercultural? | 268
 Chapter III—Summary | 273

**Summary and Conclusion** | 275

*Bibliography* | 277
*Index* | 283

# Preface

THE QUESTION OF THE origin of species is an abstract one concerning a distant past, and people can prosper without knowing the answer. However, after Darwin, it happened that precisely this question became the keystone of a number of broader and more influential ideas that affect everyone's life on a daily basis. This is why the problem ignites so much controversy and public excitement. Your attitude toward evolution does not reduce to a few highly specified biological convictions, such as whether or not random mutations can generate new functional genetic information. In fact, your position in the debate about the origin of species implies and heavily determines your attitude toward a number of other scientific, philosophical, and theological issues. What is the role of science in explaining the origins of the universe? What can necessity and chance accomplish in nature? Who is man? What is the foundation for human morality and dignity? How do you understand the Bible? How does God relate to the universe? All these and many other fundamental questions come together in the single question of origins.

Our ideas about the universe are as much connected with one another as the universe constitutes one entity. This is why convictions about one part of reality influence the perception of another. Having a coherent worldview is not just a means to living a fulfilled and peaceful life. It is an obligation and call for every human being—an obligation justified by the fact that man is a rational animal. Aristotle begins his greatest book by stating, "All men by nature desire to know."[1] Hence, "to know" means to fulfill the desire of human nature. The truth that satisfies people comes from different sources: from our senses, from our rational reflection on surrounding reality, and—not less importantly—from

---

1. Aristotle, *Metaphysics*, Book I, Part 1.

supernatural revelation by God. This last type of knowledge cannot be obtained by human endeavor, yet it is crucial for fully understanding the reality in which humans exist. This is why we cannot entertain a coherent worldview if we exclude any of the available sources of human knowledge. Instead of choosing, for whatever reason, only some facts and data while ignoring others, we need to formulate a great synthesis of faith, philosophy, and science. This book is intended to help you understand how different theories and concepts regarding the origin of species come together to constitute one coherent worldview. Surely, not all of the concepts fit equally into the picture. There are some ideas that exclude others and others that include them. Our intention is to show how some of the concepts surrounding the problem of faith and science are harmonious and why some are inherently contradictory. Thus, the objective of this work is to develop three major themes present in the debate about origins: the relationship between science and faith, the relationship between evolution and human natural knowledge, and the relationship between evolution and Christianity.

In today's debate we find a few different attempts to synthesize human knowledge based on different paradigms. The evolutionary paradigm is the one that prevails in culture. Its influence is so great that one of the ardent atheists of our times calls it "the universal solvent capable of cutting right through the heart of everything in sight."[2] I believe that the Christian paradigm greatly differs from the evolutionary one and constitutes a better background for the ultimate answer to the question of origins and all the derivative problems. Therefore, our great task is to restore the Christian paradigm in contemporary culture and to build upon it a new science-faith synthesis.

This book is designed to present the Reader with the concepts involved in the debate about the origin of species in a clear and philosophically sound way. You may discover that many of the ideas you have considered obvious and unquestionable actually leave much room for further inquiry. And furthermore, you may find that ideas which you deem obscure, irrelevant, or untenable can actually make sense. They can be put together in a broader picture that satisfactorily combines the scientific, the philosophical, and the theological level of discussion. This book provides you with a map by which to navigate between these levels. The starting-point of our journey is the current confusion in the debate

---

2. Dennett, *Darwin's Dangerous Idea*, 521.

over evolution; the intended destination is an understanding of the concepts involved in this debate and their mutual relations. This map does not make you go anywhere; it does not choose the goals for you. But it helps you to not get lost, and it invites you to choose your own position more consciously.

    Michael Chaberek, O.P., S.T.D.

# Chapter 1

# Religion and Science

This first chapter, Religion and Science, is designed to explain the mutual relations between supernatural knowledge revealed to people by God and natural knowledge that people obtain by studying nature. The relation of science and religion can be considered in two aspects—the systematic and the historical.

The systematic approach refers to the mutual dependence of science and religion outside of particular historical circumstances. Theoretically there can be several models of this relation.[1] For example, one can deny the validity of religious knowledge, reducing it to just a subjective or emotional conviction that is not rooted in any objective data or reality. Then the only source of true, objective, or useful knowledge is natural

---

1. A classic in this field remains Ian Barbour's *When Science Meets Religion*. Barbour distinguishes four positions: conflict, independence, dialogue, and integration. However, Barbour, and most authors in the field, base their models on an assumption that biological macroevolution is a "true" (i.e., testable, confirmed and established) theory explaining the origin of species. Consequently, any critique of neo-Darwinism (like intelligent design) is classified by these authors as representing the model of conflict. Given that there is no universal agreement even among scientists of the highest stature regarding the ability of neo-Darwinism to explain the origin of all biodiversity, their judgment seems premature. There are scientific reasons to not fully accept this theory; therefore, rejecting it does not mean to be in conflict with science. Barbour assumes that challenging one theory in biology (neo-Darwinism or biological macroevolution) is tantamount to challenging science as such. This is why the conflict model, as presented by him and most authors in the field, is somewhat distorted. It feeds on a confusion about the actual state of science. Throughout this book we will be avoiding this kind of confusion by showing that the field of origins primarily belongs to religion rather than science.

science. This is an approach of atheists. One variant of this kind of a reductive approach is saying that science is the matter of reality (how things are) and religion is just a matter of values (how things should be). Religion does not provide any worldview, or any description of reality; it only establishes moral norms and serves human spiritual needs.

The opposite extreme is a form of fideism, in which religious knowledge is extended to the greatest details of natural realm. According to this approach, science is a never-ending enterprise, a stock of contradictory, ever-changing theories, whereas religion gives us ultimate knowledge not only regarding the supernatural but also regarding nature. On this approach, religious belief displaces empirical knowledge based on observation and experiments.

Another possible position is "isolationism," which postulates that science and religion are two completely different realms without any common field. There is no conflict between religion and science, because divine activity revealed by religion does not leave empirically detectable marks in nature. These two types of knowledge exist on two separate levels. One example of "isolationism" is the concept of non-overlapping magisteria, which we will discuss later (see I,1,d).

Both the reductionist and the isolationist models do not seem to meet the requirements of Christianity or any religion that seriously takes into account supernatural revelation, human reason, and the order of nature. Therefore, in this first chapter I will promote a model that can be adequately called "supplementary" or "synthetic." This assumes that religion and science are neither competitive, nor disjointed, nor reducible to one another. They exist simultaneously on different levels, but they have common fields of interest and supplement each other. On this approach, the human being needs both religion and science to understand the world, himself, and God.

In Section 1 we will develop a three-level division of human knowledge which consists of theology, philosophy, and science. Thus, our binary division into religion and science gains a third level, which is philosophy. As we will show, philosophy is an independent domain that allows us to make transitions between the scientific and theological truths without falling into confusion of methods and notions. The distinction of these three levels will help us define what is naturalism (in Section 2).

The historical approach explains how the theoretical relations between faith and science have been realized throughout centuries in different historical circumstances. Therefore, in Section 3 we consider how religion, philosophy, and science have interacted throughout history. Our account includes eight stages of the development of the three domains from antiquity to postmodernity. The goal of this section is to properly balance the contribution of science and theology in answering questions of origins. For this reason, we will return to the systematic approach in order to establish the limits of science. This will allow us to see when a transition from science to philosophy and from philosophy to theology is necessary in order to answer the ultimate questions regarding the universe (Section 4).

In the history of Western thought, sometimes conflicts between scientists and theologians took place. In Section 5 we will summarize the Galileo affair. This will reveal how Galileo understood the limits of science and religion, which made room for the heliocentric model of the solar system. This and the following section (Sections 5 and 6) explain the crucial difference between Galileo's and Darwin's cases. The comparison of the theories promoted by these two famous scholars will inform us that there is no simple analogy between Galileo and Darwin. Consequently, Galileo's affair cannot be presented as a historical argument for accepting Darwinian theory of origins.

## 1. Theology, Philosophy, and Science

### a. The Three Levels of Human Knowledge

People generally may be divided into two categories, monists and dualists. Monists believe that all reality consists of just one type of being: material or spiritual. Materialistic monists say that reality consists of material being alone, whereas spiritual monists (often known as pantheists) believe that all reality is of a spiritual (i.e., immaterial) nature. The other group, dualists, believe that reality includes both types of being: material, that we detect with our senses, and immaterial, that we can approach only through our intellect. The basic assumption adopted in this chapter and the entire book is that reality is dualistic and it exists outside of the human mind. But the ways in which we know these two different realms are also different. This is why in order to know both—the entire reality—we need to employ two very different types of knowing which are called science (natural science) and theology.

What is knowledge? Roughly speaking, it is a content of a mind. It differs from the unconscious content of our mind or—more broadly—the entire content of the spiritual soul. Therefore, knowledge is something that we can reflect upon and bring to our consciousness when we need it. There are two ways of understanding knowledge. According to the first one, knowledge is the subjective content of one's mind. In this sense, we can call it *knowledge for myself*. This is all that I know, regardless of whether it is true or not, verifiable or not, accessible to other people or not. It is any content of the mind that a person is conscious of. The second understanding of knowledge deems it everything that can be made known to others by means of communication. We can call it *intersubjective knowledge*. In this sense, knowledge is everything that can be reduced to human ideas and passed on to other minds by means of a language. These two understandings of human knowledge are not two disjoint things; rather, the second may be seen as encompassing the first. This is because any intersubjective knowledge is also knowledge for myself, but not vice versa—knowledge for myself is not always intersubjective.

To make the distinction more clear, let us present a few examples. A mystical experience enriches subjective knowledge, but it is essentially incommunicable. Only some part of it might be passed on to others by means of a language. The same applies to an aesthetic experience and to intuition. They may teach a person about things, but inasmuch as they cannot be passed on to other humans, they remain just knowledge for oneself. There are types of knowledge, like secrets, that are communicable but are never communicated. In this case we can say that it is intersubjective knowledge only potentially, whereas actually it is knowledge for myself. In contrast, any knowledge about physical facts must be intersubjective. Also, any science, literature, or knowledge existing in a written form is intersubjective, because not only can it be communicated, but it already exists in a communicable form. From this point on, whatever we say in this work about human knowledge refers to the intersubjective type of knowledge.

In the middle of Diagram 1 there is a rectangle that represents the human mind and its contained knowledge. Knowledge can be divided into three domains: theology, philosophy, and science.[2] These three differ according to their subject matter (object), method, and goal.

---

2. The division at the bottom of the rectangle could include science taken in its broadest sense, which includes more than natural science. In our context, however, we leave aside some types of knowledge, such as those found in the humanities. We focus specifically on natural science, because throughout this book, the entire bottom division is geared toward discussing evolution.

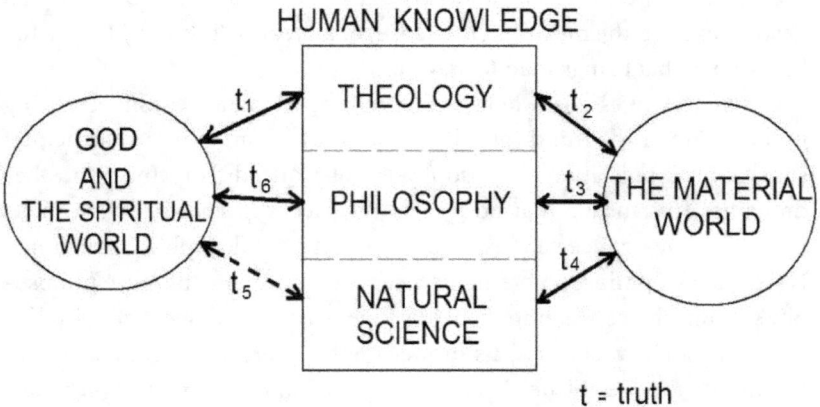

Diagram 1. The three levels of human knowledge in
the mind, as related to external reality.

There are many definitions of science, as well as of philosophy and theology. For our purpose, we need to define these three domains in a way that highlights the differences between them. Thus, we need to pin down what is essential for each domain of knowledge, as opposed to other domains. Let's begin with theology.

*Theology* consists of knowledge revealed to man in a supernatural way, either directly by God or through the mediation of the spirits, the prophets, the Apostles, the Holy Scriptures, etc. This knowledge may teach us either about the invisible reality or about the material world. God is the proper object/subject[3] of theology. Theology (in this case Christian theology) informs us about facts: God is a Trinity, God became incarnate in Jesus Christ, there are different choirs of angels, there is a hell and a heaven, and so forth. The material world is the secondary object of theology. Theological knowledge about the material world informs us about, for instance, the creation of the universe, the consummation of the world at the end of time, the final judgment, the miracles, etc. Theological knowledge may also confer many moral and spiritual truths, such as God's expectations regarding human actions or divine influence on

---

3. When Thomas Aquinas speaks about the object of theology, he prefers to use the word "subject" (*subiectum*) with regard to God, as God can hardly be an object of anything. This is why whenever we speak about God as a proper subject of theology, we will use "object/subject." (Cf. *S.Th.* I, q. 1, art. 7c.)

man through grace. The goal of theology is to give an ultimate explanation to reality (both visible and invisible) in the context of human final destination and the meaning (the sense) of all reality. Theology thus is the knowledge that brings man to salvation.

*Philosophy* is knowledge acquired by human reason reflecting upon nature or the mind (or self, i.e., self-reflection) without the help of supernatural revelation. Philosophy is not isolated from the two other domains. This means that both facts revealed supernaturally and facts discovered by science can become the object of philosophical reasoning. Nevertheless, philosophy remains a separate discipline because philosophy's main object is being as being (being *per se*), that is, being taken in the most abstract sense. Its proper method is reasoning according to the first principles of being and thought (*prima principia cognitionis*), with the use of logic and intellectual tools, such as analogy and abstraction. The philosophical method is different from the theological method because it does not adopt supernatural premises, and it is also different from science because it does not ponder particulars but rather seeks generalization on a more abstract level. The goal of philosophy is to find ultimate causes of reality by means of natural reasoning.

Finally, we have *natural science*. We have added the word "natural" to science in order to make it clear that we understand science according to the modern rather than the medieval sense of the word. In scholasticism, science (Lat. *scientia*) was considered the highest form of knowledge because it represented knowledge that was certain. This referred primarily to philosophy and theology. In the modern era, the domains of philosophy and theology were no longer considered certain because modernity considers certain only what can be empirically demonstrated. Consequently, the meaning of the word "science" changed, to signify empirical or natural sciences. Regardless of whether that shift was justifiable or not, for the sake of clarity and communicability we use the word "science" according to its modern meaning. The proper objects of science are empirical facts, and the method is observation and experimentation. Science, in contrast to philosophy, is focused on particulars and details. Its goal, therefore, is not to look for an overall picture of reality or the first cause of everything. Instead, it ponders upon the natural and proper causes of physical phenomena. Unlike theology, but similarly to philosophy, science does not resort to any supernatural knowledge (e.g., from Holy Scriptures or supernatural revelations).

We need to notice that all three domains (or disciplines) of knowledge are intertwined to some degree, and there are some "in-between" areas of inquiry that can be claimed by two or all of them. We could also identify more disciplines and subdisciplines within each domain, thereby creating more harmonious transitions between the three levels of knowledge. But this is not the important or controversial issue. Our diagram suggests something more, namely, that the disciplines are ordered in such a way that theology, being highest, seems more important than the two other domains, and science, being lowest, seems somehow subordinated to philosophy and theology.

There are at least three reasons why we should arrange the three domains in the following descending order: from theology, to philosophy, to natural science. The first reason is the dignity of the object. God is a more noble object/subject of investigation and knowledge than the created universe. Even unbelievers could agree on this, as long as they understand God in the way believers do—as an absolute and perfect Being. The object of philosophy, in turn, is being as being (being *per se*). Studying being is like studying everything, because everything that exists is being. In contrast, the object of science is only material being, and even this in a very limited aspect. Philosophy looks for the causes of all reality, whereas natural science looks for a cause of a particular phenomenon at a particular time. Consequently, philosophy finds the ultimate causes of everything, while science ends up with a number of particular causes that explain only some parts and aspects of reality. This is why we are justified in recognizing philosophy as higher than science with respect to their objects. In contrast to theology, philosophy informs us about reality only to the extent that human reason can reveal. Theology reaches God through divine revelation, which is a much more powerful source of knowledge. So, even though philosophy can possibly recognize the existence and some attributes of the perfect and immaterial Being, she is unable to reveal what exceeds natural reasoning. This "explanatory power" makes theology superior to philosophy.

The second reason to establish the descending order between the three domains is the profundity of their statements. Surely it is important to understand how the particles interact in a given compound or how amino acids are arranged in proteins, but all of this seems meaningless in comparison to death, afterlife, happiness, or salvation. Theology addresses those most profound questions that no man can shun. Philosophy also speaks about ultimate causes and ends. So, even if most people are not concerned with those issues on a daily basis, every human ultimately faces them. For example, everybody wants to be happy, and the knowledge about how to become happy belongs to philosophy and theology rather than natural science.

Finally, the third reason to place the disciplines in the descending order is the level of certitude that each of them provides. We should not confuse a degree of certitude of a given assertion with the degree of voluntary assent (Lat. *assentio*) granted by a given person to a given assertion. The former is objective and is inherent in the very assertion, whereas the latter is subjective and resides in the human will. Inherent certitude belongs only to true assertions. If an assertion is false, then it can entertain only the voluntary assent of a person holding to it, not inherent certitude. The problem is that inherent certitude does not say which theological, philosophical, or scientific proposition is true, yet it belongs only to true statements. For this reason, true and false statements may look similarly convincing. It may even happen that someone grants greater voluntary assent to an untrue scientific claim (e.g., "a majority of DNA is junk"[4]) than to a true theological claim (e.g., "God is a designer"), as well as lesser assent to a true scientific claim (e.g., "life contains irreducible complexity") than to an untrue theological claim (e.g., "there is no hell").

The degree of certitude is implied by the source of knowledge. Since the source of knowledge in theology is God, theological assertions entertain the highest degree of inherent certitude. This certitude is even higher than that of self-evident (Lat. *per se nota*) truths, such as mathematical equations. The reason is that God is never wrong and never lies, while human reason may err. This is also why theological statements are more certain than philosophical ones. When it comes to the sciences, their statements are of the least certitude because they describe changing reality and are based on empirical observations that may be mistaken. Note that this view does not lead us to support the relativistic idea of science, whereby it is just an interplay of ever-evolving postulates and a never-finished enterprise. This idea is common among some philosophers of science who are probably influenced by a broader trend of skepticism present in today's culture. Science does provide knowledge, both objective and in some cases even permanent. But the goal of science is not to

---

4. There is a widespread conviction among evolutionary biologists that the majority of DNA in the cell is a useless by-product of random evolution. However, there is mounting evidence that the so-called junk DNA actually plays roles in the cell. See Jonathan Wells, *The Myth of Junk DNA*.

provide ultimate knowledge about reality because the very reality which is the object of science is neither permanent nor ultimate. In our opinion, scientific propositions, by themselves, are reliable and convincing; their limited certitude is evident only when compared to those of philosophy and theology.

## b. Knowledge and Truth

Before we elaborate upon Diagram 1 in more detail, we need to introduce another idea—a threefold cognitive model. This can be represented in the following way:

(1) Mind—(2) Different cognitive disciplines—(3) Reality

Reality (3) is being that is objective and external to the mind. It is an object of cognition for different cognitive disciplines (2). The being may be the physical universe, God, the invisible universe, or whatever else exists in whatever form. Reality is captured by the mind (1) through the mediation of different cognitive disciplines (2), such as theology, philosophy, and science in their broadest sense. The treasury of cognitive disciplines (2) contains things like scientific postulates and theories, dogmas, adages, philosophical concepts—everything that is communicable by means of an understandable language. Through intersubjective communication, the treasures of human knowledge become accessible to all people, and all people can potentially possess the same knowledge.

In order to understand what truth is, we need to go back to Diagram 1. We see that knowledge is in the mind (represented by the middle rectangle). There are different levels of knowledge according to different cognitive disciplines, such as theology, philosophy, and science. The human mind acquires knowledge from reality. In Diagram 1, on the right there is physical reality (the visible or natural world), and on the left there is the invisible or supernatural realm, that is, God, angels, heaven, hell, etc. Both sides are connected to the human mind by arrows. This symbolizes a correspondence between mind and reality. If the mind is properly shaped by reality, then it contains true knowledge. This is a classic philosophical understanding of truth. Thomas Aquinas defines it as *adaequatio rei et intellectus*, which means that there is a correspondence between the thing itself and the mind's concept of it. In other words, truth is some kind of conformity of the mind to external things. It follows that the arrows in Diagram 1 represent only true knowledge (not any knowledge). The correspondence takes place only when the mind is properly shaped and the equality is actually present. Conversely, when knowledge in the mind does not match reality, then the knowledge is false.

The arrows in Diagram 1 point in both directions to show that even though reality is what primarily shapes the mind, there is also some influence of preconceived ideas that somehow shape our cognition of reality. Examples of this are scientific theories that fill in the gaps of our understanding of the visible world. We receive particular data and facts from reality, but to make sense of them we need theories and other generalizations. This is also why we need different cognitive disciplines that help us see one object from different perspectives.[5]

This does not mean, however, that we have no access to "things themselves." It was Immanuel Kant who proposed that the mind has no access to reality and there are some inherent categories in our minds that are necessary for making sense out of empirical perceptions. In our model, it is reality external to the mind that is recognized by the mind, and reality decides what kind of knowledge the mind contains. This is also why truth is primarily in things and only secondarily in the intellect. For Kant, categories of the intellect shape our knowledge about reality, whereas in our model categories are produced in the intellect through experience and interaction with reality in the first place. The participation of the mind in the production of knowledge does not distort reality but only helps to abstract universal ideas and make coherent notions out of perceptions. We can say that the mind uses the previous knowledge to make sense out of anything new it encounters.

Next, we need to explain why there are multiple arrows in Diagram 1. The meaning of this is pretty straightforward: there is correspondence between the two different realities (visible and invisible) on the one hand and the three levels of knowledge on the other. Reality is linked to the mind according to different levels of knowledge. Hence, we can speak of

---

5. Another example of how the mind influences our understanding of reality is the way in which we come to recognize physical things. Our sensory perception is fragmentary and we need to learn how to recognize physical objects (we learn this as infants). As adults, when we see a phenomenon of a car, we spontaneously make the judgment that it is a car. In fact, we never see a car; we see colored shapes; we may also hear the sounds and smell the exhaust of a car; but none of this is a car. Yet thanks to our earlier multiple experiences with cars, our intellect makes the judgment that we are actually dealing with a car. This is one way in which the mind corresponds to reality and pursues knowledge.

philosophical knowledge about God, theological knowledge about God, scientific knowledge about the material world, and so forth. There are, however, two types of knowledge that require some explanation. These are theological knowledge about the material world ($t_2$) and scientific knowledge about the immaterial world ($t_5$). We will address these two in the following subsections.

## c. The Two Books

There is a deep conviction in Christian tradition that God revealed himself in two ways: naturally through creation, and supernaturally through revelation. We can therefore speak of two books in which humans read the divine message: the book of nature (that is, the entire visible order), and the book known as the Bible. The first type of revelation is not as clear as the second. It speaks about God only through his works, and therefore it gives only partial knowledge that needs to be supplemented by faith. The Bible itself confirms the ability of man to know God from his creation as the Creator and ultimate cause of everything.[6] Since there are two books through which God speaks to man, there are also two truths about God and the universe; one is acquired by natural reason, and the other is revealed supernaturally and exceeds the abilities of human reason. These two truths have one and the same ultimate source, namely God. This is why they cannot contradict each other. The Catholic Church has on multiple occasions confirmed that nature and the Bible constitute two ways in which one God revealed one coherent message about Himself. Therefore, it is both against reason and against Catholic teaching to claim that knowledge from natural science or sound philosophy (*sana philosophia*) can contradict any truth belonging to faith.[7] Similarly,

---

6. Wis 13:1–9; Rom 1:21–22. This truth was later established as solemn Church teaching at the First Vatican Council.

7. This is confirmed by the Fifth Lateran Council (1512–1517), Session 8, 19 December 1513: "Since truth cannot contradict truth, we define that every statement contrary to the enlightened truth of the faith is totally false and we strictly forbid teaching otherwise to be permitted. We decree that all those who cling to erroneous statements of this kind, thus sowing heresies which are wholly condemned, should be avoided in every way and punished as detestable and odious heretics and infidels who are undermining the catholic faith. Moreover we strictly enjoin on each and every philosopher who teaches publicly in the universities or elsewhere, that when they explain or address to their audience the principles or conclusions of philosophers, where these are known to deviate from the true faith—as in the assertion of the soul's mortality

revealed truth cannot contain anything against facts of nature soundly recognized by science or philosophy. Truth cannot contradict truth—as the popular Catholic adage goes.[8] This is why, when an apparent conflict arises, a believer should assume that either theology or science is understood improperly. The believer should investigate both disciplines in order to resolve the apparent conflict.

In Diagram 1, supernatural knowledge about the physical universe is represented by arrow $t_2$, and natural knowledge about the physical universe is represented by arrows $t_3$ and $t_4$ (corresponding to philosophical and scientific knowledge, respectively). Lack of conflict means that these three truths are entirely compatible, even if practically speaking they sometimes seem to be in tension.

There is, however, some asymmetry between the competencies of science and theology. Theology provides a positive knowledge about visible reality ($t_2$), whereas science does not provide any positive knowledge about God or the invisible universe. For instance, the temporal beginning of the universe and God's works in history are known from theology, but they cannot be known by science. Similarly, science cannot say anything about God or the supernatural world except for purely negative statements. For example, when a miracle takes place, science can only say that it does not know how this event occurred based on current scientific knowledge. As another example, when science inquires into the beginning of the universe, it can go back to the initial singularity, that is, the phase immediately preceding the rapid expansion called the "Big Bang." It cannot, however, say what preceded the initial singularity. The most it can do, without exceeding its scope, is to suspend its judgment. Positive knowledge about the occurrence of a miracle or the creation of the universe out of nothing belongs to theology. For science, speaking about God is like us speaking about nothingness. We cannot think about nothing, because whatever we think of is already something. We can think about what nothingness is only by not thinking about something. Similarly, science can speak about God, immaterial things and their actions only by

---

or of there being only one soul or of the eternity of the world and other topics of this kind—they are obliged to devote their every effort to clarify for their listeners the truth of the Christian religion, to teach it by convincing arguments, so far as this is possible, and to apply themselves to the full extent of their energies to refuting and disposing of the philosophers' opposing arguments, since all the solutions are available."

8. See Pope Leo XIII, *Providentissimus Deus*, no. 23; Pope John Paul II, Address to the Pontifical Academy of Sciences, 22 October 1996; Pope Francis, *Evangelium Gaudii*, no. 243.

speaking about something that they are not. If science says anything positive about a thing, this means that the thing is neither God nor part of the invisible realm. Arrow $t_5$ in Diagram 1, representing scientific knowledge about invisible reality, is broken. This is because this knowledge has no positive content; it can only reveal the limitations of the scientific method regarding supernatural phenomena. Science speaks about immaterial reality only in a negative way, i.e., by saying what it is not.

### d. Faith and Reason vs. Faith and Science

In the contemporary debate about origins, there is confusion regarding three different intellectual planes: one of these is *faith and reason*, another is *faith and science*, and the third is *the question of origins*. Though these three are connected and even overlap each other, they should not be conflated or confused.

First, we need to realize that faith, properly understood, is rational. Faith without reason turns into fideism or fundamentalism. Neither of these is the Christian approach to faith. When St. Anselm of Canterbury speaks about "faith seeking understanding" (*fides quaerens intellectum*), he does not mean a science–faith relation, but rather a relation between human reason and divine revelation. In Christianity, faith and reason are always combined in a synthesis, and this is why we cannot isolate the two. Faith is rational, and reason—in order to know the truth—has to be enlightened by faith (*intellectus quaerens fidem*). Christianity has always considered faith and reason "two wings on which the human spirit rises to the contemplation of truth."[9]

However, it is completely justified to separate and compare the conclusions of rational faith on the one hand and the claims of different scientific disciplines on the other. This is possible because faith and science have different sources of knowledge, different methods, and different primary objects (for theology, the physical universe is only a secondary object). Hence, in matters that are common to both, they may either peacefully supplement each other or remain in tension. To resolve an apparent conflict belongs to the so-called dialogue between science and faith.

---

9. Pope John Paul II, *Fides et Ratio*, the opening sentence. Cf. Benedict XVI, Address to Members of The International Pontifical Theological Commission, December 2, 2011.

Some participants in the science-faith dialogue propose, however, that science and theology have nothing to do with each other. This position is called NOMA (non-overlapping magisteria). According to the advocates of NOMA, science and faith are so distant that it is an error to even compare them. Since there is no common ground between them, there can never be a conflict—thus affirm NOMA proponents.[10] However, if conflict is not possible, the dialogue is not real either, and a science-faith synthesis is not attainable. NOMA ignores the fact that God revealed some truths about the physical universe that cannot be known by purely human cognitive effort. These truths (at least) are intertwined with our scientific perception of the visible reality and cause the two magisteria to overlap. Indeed, as is shown in Diagram 1, natural science and theology do have a common object, namely, the physical universe. God not only revealed some things about the visible order ($t_2$); he also works supernaturally in the visible order, and consequently some events in the visible order are inexplicable without resorting to theology. NOMA, therefore, is not compatible with the Christian understanding of the relation between faith and science. For nonbelievers, however, NOMA is an easy way to remove faith from any rational investigation and relegate it to the realm of mythology, emotions, or values alone. This is why NOMA proponents usually end up in scientism, which claims that only science gives useful, objective, or true knowledge, whereas religion is reduced to morality and/or subjective, personal convictions that can never claim any universality. Ultimately, faith appears irrational because rational investigations can take place only in science. It is true that if all faith were irrational then the dialogue between faith and science would boil down to the dialogue between faith and reason. Some NOMA proponents actually believe that this is the case, and this is why they confuse the faith-science dialogue with the faith-reason dialogue.

Both faith and science may address the question of origins (of the universe, species, and the human race) while remaining entirely within their own fields and using their own methods. This is why the dialogue between faith and science is not the same as the *faith and origins* or *science and origins* problem.

There are only two principal answers to the question of origins—creation or evolution. Hence, the possible problems are the compatibility of *faith and evolution* and the compatibility of *science and creation*.

---

10. S. J. Gould, *Rocks of Ages: Science and Religion in the Fullness of Life.*

These problems, however, are not identical to the dialogue between *faith and science* as such; they constitute just two out of many possible issues within this dialogue. Those who confuse the science-faith dialogue on one hand and the problem of origins on the other typically believe that faith has nothing decisive to say about origins. They usually claim that faith tells us only *that* things were created or *that* the universe ultimately comes from God, but it does not tell us anything about *how* things were created or formed. They deem this latter question to belong entirely to science. If this were the case, the question of origins would be settled by science, and after science had established how things began to exist, the only problem to resolve would be the compatibility between science and faith. Then, indeed, the issue of faith and origins would boil down to the dialogue between faith and science. In our opinion, however, this approach is not congruent with a Christian understanding of the role of faith in explaining the question of origins. In fact, it is faith (not science) that gives us positive and ultimate knowledge regarding the question of origins. Diagram 2 organizes the different planes of this debate.

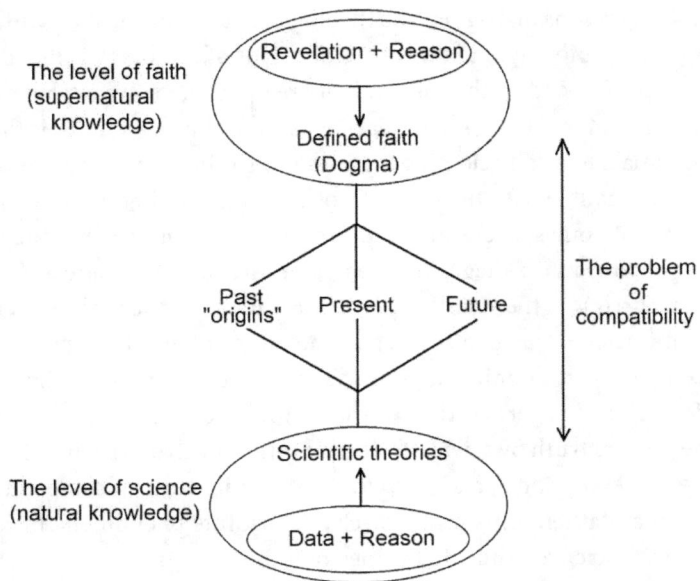

Diagram 2. The relation of faith and science and the problem of compatibility.

We see that both science and faith include reason. Hence, there cannot be incompatibility between faith and reason or science and reason unless the faith is fideism or the science is some ideology imposed on science. The problem of compatibility applies primarily to the vertical movement in the diagram. For example, we can legitimately ask if a theological concept is compatible with revelation. If not, it is called a heresy—an untenable theological idea. Established theological claims are dogmas in a broad sense. Similarly, we can ask if scientific theories are compatible with scientific data. It may happen that a given theory imposes something on data rather than derives its conclusions from data. In that case, it is more ideology than science.

Both science and faith can address three types of questions—regarding: (1) the past (such as the origin of the physical universe, planetary systems, natural history, the origin of life, the origin of species, the origin of the human race); (2) the current operation of the universe; and (3) the future of the universe.

In science, questions regarding the past are addressed in disciplines such as history, geology, archeology, and cosmology. One good example of a theory of origins proposed in science is the theory of biological macroevolution, which is intended to explain the origin of species. Faith also addresses questions of origins. The theological discipline dealing with the past is called protology, and one of the divisions of protology is the theology of creation. Within the theology of creation, there are explanations for the origin of the universe, life, species, and the human race. It is therefore legitimate to ask if scientific theories such as biological macroevolution are compatible with the theology of creation and dogmas of faith.

When it comes to questions regarding the present state of the universe ("present" in the diagram), theology postulates that there are some events occurring in the visible order that are inexplicable within science, such as miracles. If the theological truth about miracles is to be preserved, science needs to recognize its limits. The theological truth is that God occasionally acts supernaturally in the natural order. The recognition of this theological truth by scientists does not ruin science as such. Miracles are rare and exceptional, and they do not occur in order to cause natural phenomena. Rather, miracles themselves constitute phenomena inexplicable within science. Thus, faith does not enter the domain of science. It does not ruin scientific naturalism. It only claims that science cannot explain everything, and thus scientific naturalism extends only as far as the workings of nature are possible.

When it comes to the future of the universe, science predicts the uniform operation of the laws of nature, until they collapse owing to the disintegration of the physical order and the ultimate heat-death of the universe. Faith has also its own predictions, in the discipline called eschatology. Faith predicts that before heat-death occurs, God will finish human history and establish a new order of creation. We see, therefore, that in all three aspects (the past, the present, and the future), science and faith need harmonization. They may be interpreted in such a way as to exclude each other, or to supplement and enrich each other. The Christian approach strongly favors the complementary character of both, to the point that Christians are invited to build a science-faith synthesis.

## 2. Naturalism

After establishing the general picture of different types of knowledge (Diagram 1), we are ready to answer the following question: What is naturalism? Naturalism can have two general forms. The first is an attempt to exceed the confines of natural knowledge regarding God and the invisible universe (i.e., extending $t_5$ in Diagram 1). The second is a denial of supernatural knowledge regarding the visible universe (i.e., diminishing $t_2$). The first type we can call naturalism "by expansion," because its goal is to expand science beyond its scope. The second type can be called naturalism "by rejection," because it rejects theological knowledge about the universe. Rationalism, according to Diagram 1, would be expanding philosophical knowledge about God and the invisible realm beyond its limits (expansion of $t_6$).

### a. Naturalism by Expansion

According to Diagram 1, naturalism by expansion attempts to transform the broken line $t_5$ into a solid line, thereby allowing science to provide positive knowledge about God. Obviously, what is invisible and immaterial cannot be scrutinized by the scientific method, which, by its very nature, can deal only with physical phenomena. Therefore, if one attempts to investigate God (or the invisible realm) using scientific tools, one must first assume that God is actually a part of the visible universe or is simply nonexistent. In the first case, one ends up in monism. The second case takes place when a scientist says something like, "There is no God, because we cannot detect him," or, "Science has proven that God does not exist."

According to naturalism by expansion, the scientific method has no limits—everything that exists is the subject of science, and if anything is beyond science it simply does not exist, at least not in reality. This assumption, however, leads to circular reasoning. For example, naturalists say that there is no soul because we cannot empirically detect it, but their conviction stems from the assumption that everything that exists must be empirically detectable. The naturalists' conclusion is embedded in the assumption, and thus it does not provide any knowledge about the existence of the soul, God, or any non-empirical being.

Is atheism necessarily implied by naturalism by expansion? The answer depends on how far one stretches science beyond its actual capacity. If naturalism by expansion means that everything that exists may be investigated by science in a positive way, then it implies atheism (because God is beyond the scientific method) and materialism (because any immaterial reality is beyond the scientific method). But naturalism by expansion may also mean that the only true knowledge is that which comes from science, whereas anything that is not covered by science is either imaginative or useless or irrelevant. In this sense, naturalism rejects other-than-scientific forms of knowledge, but it does not imply that whatever is not the subject of science does not exist. In this moderate form, naturalism allows that there may be some objects beyond the sight of science—however, either we cannot know anything about them, or knowledge about them is worthless. If a scientist adopts this kind of thinking, then he ends up close to agnosticism. Naturalism by expansion implies either atheism and materialism or agnosticism, depending on the intention of the naturalist extending the scientific method beyond its actual scope.

### b. Naturalism by Rejection

We said above that the secondary object of theology is the material world. This means that there is some knowledge revealed to humans supernaturally by God that concerns not God but the physical universe (arrow $t_2$ in Diagram 1). This type of knowledge can be neither obtained by natural (whether scientific or philosophical) investigations nor invalidated by them. This is because theological claims go beyond what human reason is able to learn on its own. By their very nature, theological truths reveal those things that humans could not apprehend without supernatural revelation.

There are two distinctive epochs in the history of the visible universe. One is the history of creation, i.e., everything that happened from the creation of the universe out of nothing (in Latin, *creatio ex nihilo*) until the creation of man. The other is the history of salvation, which began with original sin and will last until the consummation of the world, the final judgment, and the glorious coming of the kingdom of Christ. Accordingly, there are two types of supernatural knowledge about visible reality ($t_2$). One type concerns the history of creation, the other the history of salvation.

To secure a better understanding of both categories, we will provide a few examples of truths belonging to each one. Among the truths ($t_2$) concerning the history of creation are the creation of the universe out of nothing at the beginning of time, the supernatural formation of the universe within the time period described by the Bible as the six days, the supernatural formation of animals out of the ground (Gen 2:19), the supernatural formation of the first human body (Gen 2:7), and the formation of the first woman from Adam's rib (Gen 2:22). Among the truths revealed by God concerning the history of salvation are the prophecies of both covenants regarding the future, miracles such as healings, deflections of planetary movements (e.g., Josh 10:13, Matt 24:29), and multiplications of food (1 Kgs 17:16, Matt 14:15ff.), and supernatural revelations of the Holy Spirit. All of these are truths revealed to man by God that concern the events in the visible order and cannot be known by scientific (i.e., natural) investigations alone.

Naturalism by rejection may be stronger or weaker depending on how much of the truth $t_2$ is rejected. We can distinguish between first- and second-grade naturalism. First-grade naturalism is the rejection of the majority of the truth regarding the history of creation. Second-grade naturalism is present when also some revealed truths of the history of salvation are rejected. A good example of first-grade naturalism is the rejection of the special formation of the human body or the supernatural formation of species. In fact, a majority of contemporary theologians accept *creatio ex nihilo* (the first act of creation of the universe) but reject any direct or supernatural formation of the world by God, instead assuming that different forms of life came about through natural evolution.

Proponents of second-grade naturalism typically challenge the supernatural character of miracles (all or just some of them); they may go even as far as the denial of the Resurrection or the Incarnation of Christ. Theoretically, a naturalist who rejects the revealed truth about the history of salvation could still accept the truth about the history of creation, and *vice versa*. In our times, however, the denial of the supernatural events of the history of creation with a simultaneous acceptance of the supernatural events of the history of salvation is quite common, whereas the opposite approach is extremely rare if not nonexistent. Apparently, for contemporary people, it is more difficult to accept the biblical message about creation than the biblical message about salvation. Hence, the way in which naturalism progresses in our times is from the rejection of the supernatural history of creation to the denial of miracles and supernatural history of salvation. This is why, in our typology, second-grade naturalism consists of first-grade naturalism extended by the rejection of supernatural knowledge about the salvific events revealed in the Bible.

As we said above, naturalism by expansion implies atheism or at least agnosticism. Interestingly enough, naturalism by rejection has been adopted by many Christians, both theologians and scientists, regardless of denomination. Naturalists of this kind usually accept the theological truth about God ($t_1$). Thus, naturalism by rejection does not imply atheism or agnosticism. It does, however, work against a coherent commitment to Christian truth. Diagram 3 summarizes our typology of naturalism.

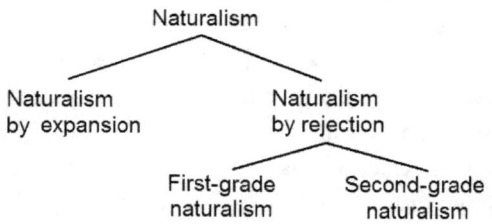

**Diagram 3. Types of naturalism.**

## 3. Historical Interdependence between Religion, Philosophy, and Science

### a. The Historiosophy of Auguste Comte

Having presented the basic systematic distinctions and relations between theology, philosophy, and science, we will now approach the same problem from the historical perspective. A starting point is the historiosophy of French philosopher Auguste Comte (1798–1857).

Comte thought that positive knowledge is all about facts—though not any facts, but only physical facts—and that only positive knowledge has any value for men. Any speculation or abstraction blurs people's minds and should be avoided. He was convinced that physical facts are the only ones worth studying, because only they make us resistant to fantasizing and deception. Studying only positive facts ensures that our knowledge is true, and by "true" he means realistic, certain, and rigorous. Comte believed that philosophy should avoid mystery and curiosity and focus on pragmatic knowledge, namely, knowledge that is conclusive, objective, and useful. Comte's positivistic mind-set led him to reject psychology, but the core of his criticism was aimed at theology and metaphysics. According to Comte, the predominance of these two disciplines was a major obstacle to the development of humanity throughout history.

Comte is an important figure of the past because he was the first modern thinker to define and encourage the pragmatic and minimalistic approach to philosophy that prevails into our own times. According to Comte, humanity has undergone three stages: the first was theological, the second was metaphysical, and only the last was positive. This is the fundamental law of humanity's intellectual progress. In the first stage, humanity explained phenomena by recourse to spirits; in the second phase, by recourse to abstract notions. In the first stage humanity was driven by emotions; in the second, by the intellect. Nonetheless, both emotions and intellect produce fiction. Emotions produce the fiction of mythologies, whereas the intellect produces the fiction of metaphysics. Mythology and metaphysics subsequently dominated humanity, and only after humanity had been freed from them could it enter the last and highest stage of development, which is positivism.

Comte, similarly to Hegel and a few other "big narrators" of the nineteenth century, sees the greatest achievement of humanity in his own doctrine. He speaks about the three stages of human progress, but in fact the first (theological) and the second (metaphysical) were similarly delusive. The middle stage was closer to the first than to the last; indeed, according to Comte, metaphysics is a kind of theology and theology does not differ much from mythology. This is why both disciplines should be equally abandoned. In his historiosophy there is a kind of developmental necessity that is as unavoidable as it is totalitarian. It takes humanity from its infancy of religion and theology through the youth of philosophy and metaphysics to the adulthood of science. The previous forms of knowledge are doomed to disappear, while the only true knowledge—science—emerges, grows, and finally takes over the totality of the human mind.

Today, we can see that humanity did not stop at the positive phase, as Comte imagined. Instead, it continued from positivistic reductionism through the optimism of modernism to the relativism of postmodernity. In recent times, postmodern philosophers have announced the collapse of the "great narrations," among which, ironically, we find also that of Comte. Yet, despite postmodern pluralism and confusion, there is one thread in Comte's philosophy that proved quite persistent over the decades. In this sense, the French philosopher is still alive. It is the idea that science is the only true knowledge, which is by default hostile to religion and must replace it. We do not see this happening, but we constantly see people who think along Comte's lines and work hard to make his ideas real. Some of them put great effort into propagating science as a tool of salvation for humanity. In their minds, science is the source of ultimate happiness and thus displaces religion.

Comte's historiosophy is more like a description of what he desired to be true rather than what was actually true. He wanted humanity to undergo the kind of transformation he described, but even in his times it was clear that this was not what had actually been occurring. Nevertheless, the question of the historical interdependence between theology, philosophy, and science is a valid and important one. In the following subsection we will describe how the relations between the three domains of knowledge actually unfolded throughout recorded history. We will see that they underwent more complex evolution than Comte imagined. This will build a historical context for our subsequent study of the conflicts between science and religion that have taken place in the modern era.

## b. Science, Philosophy, and Religion throughout the Ages

The following diagrams, 4A through 4H, represent eight stages of science-philosophy-religion interdependence from prehistoric times to the present day. Each of them captures a peculiarity of the relations between faith, philosophy, and science at a given historical moment. Since theology as an academic discipline could be too narrow a notion to properly correspond with the diagrams, we use here the terms theology, faith, and religion interchangeably. In this way we refer to a broader reality of revelation rather than merely formal Christian doctrine.

The circles in the diagrams represent the predominance of a given discipline at a particular time. The larger the circle, the greater authority a given discipline enjoys in the mainstream culture of its time. If the circle is smaller, the discipline is less important. The largest circle corresponds to a discipline that encompasses all other human intellectual activities. People tend to grant more power, authority, and competence to this discipline than to the two others. For instance, people derive from the discipline an answer to ultimate questions, such as the beginning and end of the universe, the meaning of human life, human happiness, and salvation.

Each of the eight stages depicts the mainstream approach present in Western civilization at a given time. It does not mean that the diagrams have no exceptions. If, for example, at some time science becomes dominant, this does not mean that every single person attributes more power and authority to science than to religion or philosophy. Rather, the diagrams present the general spirit of a given epoch, not cultural niches or the exceptionality of some individuals. Hence, our diagrams simplify reality in order to show only the mainstream tendencies of the evolving culture.

## 1. Primitive Peoples

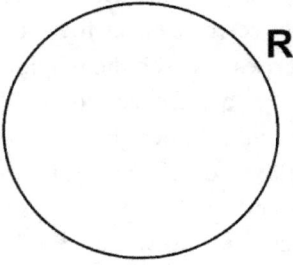

Diagram 4A. Primitive peoples.

Prehistoric peoples did not pursue science or philosophy in the intersubjective sense (see our previous discussion concerning the meaning of *intersubjective knowledge* in I,2,a). The life of primitive peoples extends from the simple crafts of daily life to religious interpretation of the surrounding world. Religion provides answers to all the important questions. Hence, in Diagram 4A, the all-encompassing discipline is religion.

## 2. Pagan Antiquity

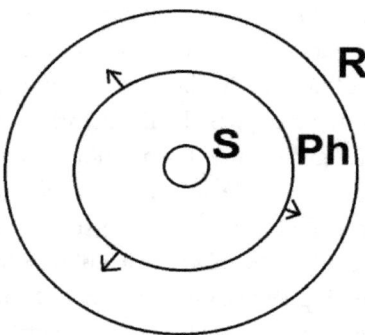

**Diagram 4B. Pagan antiquity.**

In ancient Greece, philosophy emerges. Philosophy begins with a rational reflection on reality and the search for the *Archē*—the principle and first cause of everything. Mythology, with its deficient gods and contradictory stories, does not seem to meet the intellectual expectations of philosophers. Tension arises between the religious and the philosophical interpretation of reality. Plato and even more so Aristotle do not see mythology as the ultimate truth. Rather, they see its limitations and tend to reduce religion to morality and public duty. Philosophers' reason presses on the religious attitude and subdues it. Perhaps we see this conflict most clearly in the life of Socrates, who was a great humanist and a philosopher finally sentenced to death for corrupting the minds of the youth and disregarding the gods.

Romans, after invading Greece, created their own mythology and later also philosophy—maybe less metaphysical and abstract, more ethical and pragmatic, but still philosophy. Philosophy attempts to subdue the myths, but philosophy is for the elites. Lower classes are satisfied with the myths. In Diagram 4B, religion still has the leading role in culture, but it is definitely oppressed by philosophy, which tries to rationalize it and displace it. The arrows represent this pressure of philosophy imposed on mythological beliefs. The small circle in the middle represents the emergence and slow development of sciences, at the time mainly astronomy and mathematics.

## 3. Early Christianity

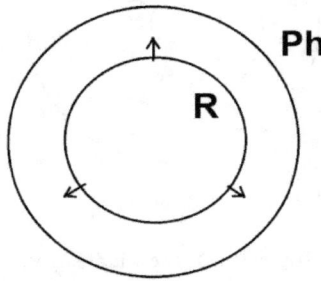

**Diagram 4C. Early Christianity.**

Christianity brings new faith and new knowledge about God, man, and the universe. Early Christians encounter philosophy, and they need to address its claims. As Richard Niebuhr shows,[11] there are different approaches to philosophy—some Christians reject it (e.g., Tertullian, St. Clement of Rome), while others endorse it and try to make use of it (e.g., St. Justin Martyr, St. Clement of Alexandria). But whichever position Christians adopt, they need to free philosophy from whatever is not in accord with the faith. In this sense, a tension between religion and philosophy arises. The more power the Christian religion gains, the more is pagan philosophy marginalized. This trend culminates with the shutdown of the Platonic Academy in Athens (AD 529) which was the longest-lasting school to teach pagan philosophy. Some great Greek philosophers fall into oblivion for centuries, when they suddenly reemerge in the Middle Ages after their works are translated into Latin. Yet, even then, Aristotle begins his new career in the West with condemnation from Church officials. We see, therefore, that in the first millennium theology

---

11. H. Richard Niebuhr, *Christ and Culture*.

surpasses philosophy and becomes the leading power in culture. There is minimal progress in science during this time.

## 4. The Middle Ages

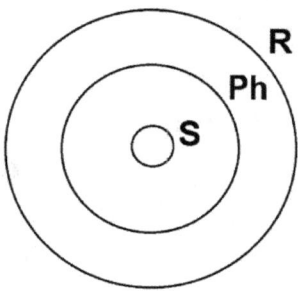

**Diagram 4D. The Middle Ages.**

Now the time of a synthesis comes. In the Middle Ages, religion is too strong and too well established to be afraid of the encounter with philosophy. The condemnation of Aristotle is overturned by the work of Thomas Aquinas. The medieval doctors of the Church include in their syntheses the teachings of the two greatest pagan philosophers (Aristotle and Plato). Still, theology is the queen of the cognitive disciplines. All conclusions of philosophy remain in accordance with the demands of the faith. Different authors write the *summae*—books designed to present synthesis of faith and reason, theology and philosophy, the *revelatum* and the *revelabile*. The rediscovery of Aristotle is accompanied by the re-emergence of natural science. Ancient science had been preserved in the Arabic world and was introduced to the West, along with the philosophy of Avicenna and Averroes. Medieval science consisted mostly of commentaries on natural works by Aristotle. In this sense, science remained a part of philosophy, namely, a natural philosophy (*philosophia naturalis*), entirely dependent on the higher philosophical principles. Experimental science was definitely not the driving force for medieval culture. Some experiments were conducted, but the substance of knowledge and the basic physical theories seem to be entirely inherited from antiquity. Medieval scholars did not question Ptolemaic astronomy or Aristotelian physics.

Instead, they considered the ancient scientists as authorities on natural matters. The Middle Ages represent a harmonious interdependence of the three disciplines: theology is the queen of all sciences, philosophy is a help to theology, and natural science is a child in the womb of philosophy.

## 5. Early Modernity

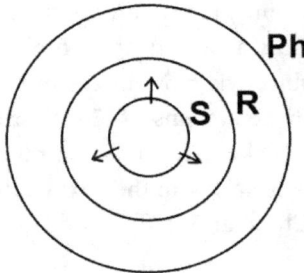

**Diagram 4E. Early Modernity.**

A number of events lead to the destruction of the medieval synthesis and probably the greatest cultural transition in history—from the Middle Ages to modernity. The Protestant Reformation disintegrates the political unity of Europe referred to as Christendom (Lat. *Christianitas*). The rejection of papal and Church authority, along with divisions regarding doctrinal issues, ends the period of theological uniformity. Philosophy, on the other hand, with its rational method and apparently certain conclusions, becomes more appealing. Since religion causes divisions and fights, those who want to present universally valid truths leave religion aside and turn to philosophy. In this way, the detachment of philosophy from theology begins—the phenomenon observed up to the present day. Once philosophy rejects the light of theology, reason takes its own course and soon begins to wander between multiple contradictory systems.

William of Ockham and John Buridan (scholars of the early fourteenth century) introduce nominalism and initiate the destruction of the abstract and analogical thinking that is crucial for understanding classical metaphysics. René Descartes (1596–1650) attempts to discover an absolute knowledge that would be attainable by the power of the rational method alone. The starting point for Descartes differs from that adopted by Aristotle or Aquinas. The latter two begin with perception of external reality, and this produces truth that consists of establishing equality between the mind and external things. In contrast, Descartes believes that the closer to the mind the object of cognition is, the more certain the cognition is. Hence, the most certain is the cognition of things contained in the mind itself. What is clear and distinct is also certain, and only certain knowledge is true knowledge. The classical understanding of truth as equality between reality and the mind is replaced by a new criterion consisting of clarity and certainty. From now on, the emphasis in philosophy is on the subject—the human mind, the "thinking I," and thus begins the era of prevailing subjectivism. Mainstream philosophy is shaped by the subsequent rationalistic systems of Hugo Grotius, Baruch Spinoza, Thomas Hobbes, John Locke, Nicolas Malebranche, and Gottfried W. Leibniz, which will have offspring in the next (Enlightenment) period in the thought of Johann Christian Wolff, David Hume, Jean-Jacques Rousseau, and Immanuel Kant.

In Diagram 4E, philosophy is represented by the largest circle. This means that philosophy becomes the encompassing discipline and the driving force in culture. In modernity, mainstream culture is subordinated to philosophy, which attempts to rule over religion, changing its meaning and content. European thought plunges into rationalism. Throughout the sixteenth and seventeenth centuries, the unleashed reason tries to find its own domain, a domain independent from theology. Ultimately, however, reason ends up in hostility toward religion and attempts to replace it with its own systems of thought. Reason is divinized, detached from faith and worshiped as an idol. The climax of this tendency is clearly seen in the French Revolution, which ends up establishing the cult of Reason as a state religion.

During the transition from the Middle Ages to modernity, another shift occurs involving pressure exerted on theology by science. The discovery of the New World, the disproval of the Aristotelian idea of the perfection of heavenly bodies (the sun has dark spots, the moon's surface is not even), and especially the heliocentric theory proposed by Nicolaus Copernicus—these and other developments shook the foundations of the medieval worldview. The controversy over heliocentrism, with its peak in the Galileo affair, and the overall growth of scientism in European culture are represented by the arrows originating in science and pressing upon religion. In modernity, science is born from philosophy's womb and slowly becomes an independent realm.

## 6. Enlightenment

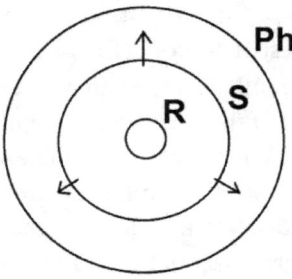

**Diagram 4F. Enlightenment.**

The mainstream culture of the Enlightenment reduces philosophy to a kind of a play with different concepts, principles, and systems of thought. Philosophy turns into philosophizing and reasoning into rationalism. This entails detachment of reason from theology and the blind trust in the power of deductive method. Multiple idealistic systems arise in Germany, whereas in France philosophy attempts to replace the Christian religion with the religion of reason. The ruling party, as in antiquity, now again consists of philosophers. In Diagram 4F, philosophy remains the largest circle. Parallel to the bloom of rationalism is the flourishing of the experimental method and mathematical description in science. The arrows in Diagram 4F show the pressure exerted by science upon philosophy. This refers to the attempt of science to take over the leading role in culture.

One example is the work of the French *encyclopédistes*, who want to bring together all human knowledge and show its importance for society, education, and progress. The extended title of their multivolume work reads: "Encyclopedia or Reasoned Dictionary of Sciences, Arts, and Crafts." The title conveys a clear message: reason has a leading role, and knowledge consists first and foremost of the sciences. Theology and philosophy are not even mentioned. One of the strongest ideas of the

Enlightenment is *deism*, which is a philosophical system that draws heavily on scientific discoveries. An enchantment with the uniformity and explanatory powers of the laws of nature leads many scientists and philosophers to the conclusion that God was needed to create the universe and establish the laws of nature, but after initiating the material realm is not needed anymore, natural laws and forces governed by mathematical principles being sufficient do the job. Miracles and extraordinary providence do not happen, according to deists. Isaac Newton shows that angels do not need to move celestial bodies in their orbits—the forces of gravity are sufficient. Pierre-Simon Laplace, when asked by Napoleon about the role of God in his model of the solar system, replies with the famous *"Je n'ai pas eu besoin de cette hypothèse"* (I had no need of that hypothesis).[12]

Another contribution to this shift is the birth of a new discipline—geology. In 1788, a naturalist, James Hutton, takes two companions on a boat trip along the eastern coast of Scotland. They land at Siccar Point, where Hutton demonstrates palpable evidence of a very long history of rock formation. Hutton concludes, "We find no vestige of a beginning, no prospect of an end."[13] The discovery challenges the long-maintained convictions that the earth is six thousand years old and that the universe was formed within six natural days. The concept of "deep time," i.e., the idea that the timescale of the universe spans not thousands but millions and billions of years, creates tension between the naturalists and religious circles that are eager to defend calculations based on biblical records. A few decades later, Charles Lyell proposes that geological formations are not attributable to any supernatural actions of God but, instead, are formed according to uniform physical laws working over vast periods of time. Around the same time, a few naturalists, among them Erasmus Darwin, come up with the idea that species of animals and plants were not created separately but evolved from "one living filament" thanks to the "power of generation." In Erasmus Darwin's thought, God remains the Creator and the great Architect of the universe, but He does not create the effects. Instead, he creates the causes of effects, and this reveals His magnificence and infinite power. During the Enlightenment, rationalistic philosophy

---

12. There is some debate among the historians whether Laplace ever said anything like this or not, but the very fact that the story appeared around that time reveals the spirit and the general tenor of the scientific debates of the Enlightenment. See https://en.wikipedia.org/wiki/Pierre-Simon_Laplace (29 March 2016).

13. "The result, therefore, of this physical inquiry is, that we find no vestige of a beginning,—no prospect of an end." Hutton, *Theory of the Earth*, Vol. 1, 200.

is the leader, and science emancipates the world from philosophy while constricting religion.

## 7. Positivism

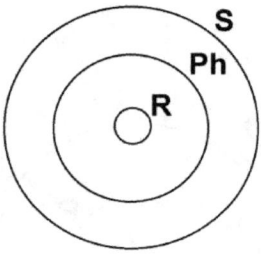

Diagram 4G. Positivism.

We have already shown how Auguste Comte considered science the ultimate knowledge of humanity. According to his system, science is to replace religion, and philosophy's task is reduced to organizing the conclusions of different scientific disciplines. Philosophy is to be merely a methodology of sciences. Religion, in turn, is reduced to myths and fairy tales. Positivism abounds in different "-isms" that can all be covered by one term: "reductionism." Reductionism is a philosophy that finds an explanation of a higher phenomenon in a lower cause. In positivism we see religion explained by a psychological complex (Freud) or a resentment feeding the hearts of people who are unsuccessful (Nietzsche); we also see religion reduced to a delusion that the rich offered the poor in order to exploit them (Marx). In all these systems, religion is explained in immanent terms; it never has any higher or supernatural source.

Charles Darwin perfectly fits into the time of "-isms." He is quite explicit about the goal of his theory, namely, to remove the continuous "interventions" of the first cause.[14] Thus, in Darwin's theory supernatural causation is reduced to merely a play of chance and necessity according to the laws of nature. Darwinism, along with Marxism, Freudianism, Nietzscheanism, etc., is aimed at minimizing the role of religion in culture

---

14. Darwin, *The Descent of Man*, 152–53.

and subordinating religion to science. We see in Diagram 4G that the positivistic understanding of relations between science and religion is exactly opposite to the medieval one. The spirit of positivism has been strong since the mid-nineteenth century, and even now it remains a vital—if not predominant—power in culture.

## 8. Postmodernity

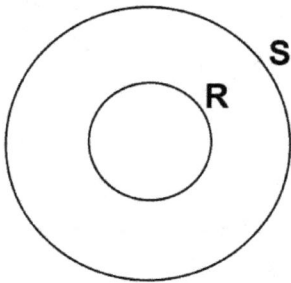

**Diagram 4H. Postmodernity.**

Diagram 4H, corresponding to the postmodern era, is characterized by the absence of the circle representing philosophy. In our opinion, the very principles of postmodern philosophizing make philosophy impossible. Philosophy (as the word itself points out) is the love of wisdom. Wisdom, in turn, is knowledge of ultimate causes and principles. Hence, philosophy is sometimes defined as the search for the *Archē*, the principle and first cause of everything. In contemporary philosophizing, minimalism and skepticism are the default positions. This goes directly against the very meaning of philosophy, and for this reason the postmodern narration can hardly be called philosophy.

Twentieth-century thought flows through two major channels: one, the so-called analytic philosophy, which gains popularity mostly in the English-speaking world; the other, the existential type of narration, which is dominant in France and elsewhere on the Continent (e.g., Sartre's existentialism, socialism, philosophy of dialogue). The analytic type, with its focus on the mind and logic, might be vaguely associated with the great Aristotelian-Thomistic tradition of European thought, whereas the existential type (with its focus on society, life, and the human will) vaguely resembles the Platonic-Augustinian tradition. Both types are generally uninterested in addressing any of the great philosophical questions.

The minimalistic approach in the early analytic philosophy (seen even more clearly in logical positivism) consists of nominalism, reduction of metaphysics to logic, and a focus on language and definitions. This type of analytic philosophy is unable to raise important questions, because the task of asking the question itself gets mired down in analyses and deductions whose only goal is to make the very question possible. Consequently, this type of philosophizing never becomes philosophy according to its original meaning.

In recent decades, the existential type of philosophizing has been generally abandoned for "deconstructionism," which proclaims the deconstruction and death of all "great narrations." Yet, if none of the great narrations can claim the truth, then the postmodern one cannot either. Hence, by its own principles, postmodernism makes itself irrelevant and meaningless. This kind of philosophy demonstrates its failure by the rejection of things like universal experience, common knowledge, and common sense. Since in postmodernism all terms and concepts are deprived of their natural meaning, they must be defined each time from scratch. In effect, no concept is understandable on its own terms, and no truth can be considered as objective. Philosophy's task is not to look for comprehensive answers but to play with language and meanings—philosophy is merely a play of words. Consequently, postmodern philosophy either reduces philosophy to literature (no wonder so many novelists claim to be postmodern philosophers, and vice versa) or commits philosophy unreservedly to one of the political ideologies. In the latter case, philosophy loses its autonomy and becomes just a tool used by a social party to gain popularity or promote a worldview.

The absence of the philosophy circle in Diagram 4H does not mean that in our times there are no philosophy departments, or people who claim to be philosophers, or those who make their living by publishing books with "philosophy" in the title. It simply means that mainstream philosophy does not play the same role that it used to play throughout previous centuries (regardless of its brighter and darker moments). It also does not mean that there is no deep reflection in our times. On the contrary, our era has witnessed a great revival of Thomistic, patristic, and Augustinian thought, as well as some great new syntheses, such as realistic phenomenology, personalism, and philosophical anthropology (including what has become known as the theology of the body). Another interesting insight may come from analytic philosophical theology. Still, all these deep and comprehensive reflections exist in academic niches rather than mainstream culture. We can say that, in postmodernity, pursuing philosophy that addresses the great questions—such as, Who is God? Who is man? Where does the universe come from and where is it headed? What is the ultimate goal of human life? What is happiness?—has become a countercultural activity.

In Diagram 4H, the circle representing science encompasses the circle of religion. This symbolizes the unprecedented extent of cultural authority that natural science has gained in our times. This circle also means the expansion of naturalism through science, a process manifested, for example, in the emergence and growing popularity of the multiverse hypothesis, the unfading power of Darwinism in popular culture, and the use of biology and physics to disprove creation and design.

## 4. The Limits of Natural Explanations

Does science have limits? If so, is there any way to establish them? Or maybe science progresses infinitely, and even if there are limits, we cannot know them? Perhaps, even if science cannot explain something now, it will be able to explain it in some distant future?

In the previous section we showed how the authority of science and its role in culture changed over the centuries. Now we will switch to the systematic mode to address the question of the ultimate limits of science in explaining reality.

Generally speaking, science is unable to find its own limits using just its own method. *Nemo iudex in causa sua.* Searching for the limits of science using the scientific method would be something like an attempt to investigate a pair of glasses by looking through the same pair of glasses, or repair a tool using the very tool. Science needs an external source of knowledge in order to establish its own limits. As shown in Diagram 1, there is an order of the disciplines of knowledge that makes it possible to judge a lower discipline in light of a higher one, as long as they speak about the same object (as with $t_2$, $t_3$, and $t_4$). When looking for the limits of science, we need to resort to higher forms of knowledge—philosophy and theology.

There are, however, some areas in which science, resorting to its own method, can know the very limits of science in the same way as we can learn something about a pair of glasses while looking through them. For example, we can learn about their transparency or how much further we can see when using the glasses. Similarly, science has certain ways of establishing how far it can go. In what follows (I,4), we will present some instances of the limits of science recognizable within science itself. Later (I,6; III,2) we will show how theology indicates other limits of natural science.

## a. The Limits of Science in Terms of Space

Since prehistoric times the scope of natural knowledge (science) has been growing. People have been inquiring into an increasingly broad physical reality. They have been curious as to what underlies the structures we observe and what is beyond what we can observe. Diagram 5 represents the tentative limits of science in terms of space.

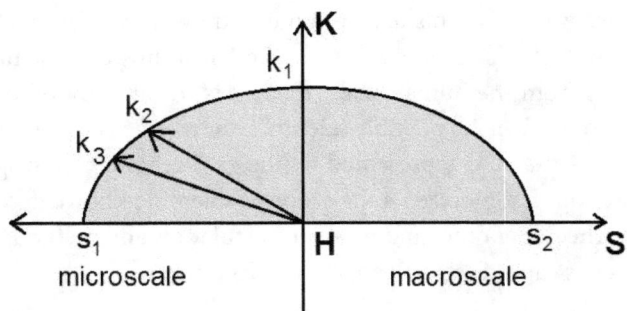

**Diagram 5. Scientific knowledge with regard to space.**

Point H represents something we could call the "human scale." This is the scale of activity and inquiry for primitive peoples. The human scale may be understood through reference to daily experience. We normally do not deal with objects smaller than 4 millimeters, which is roughly the size of letters in print. Someone who works in fine crafts may be dealing with objects on the order of 1 or 0.5 of a millimeter (such as the heads of pins or very fine threads). When objects are smaller than this, we employ tools, such as a magnifying glass or a microscope. The human scale is the scope of human activity in which we do not normally resort to observation tools. When we need a magnifying glass or a microscope, then we move from the human scale toward the microscale (that is, from point H toward point $s_1$ in Diagram 5). The smaller the objects, the closer to $s_1$ we move.

A similar thing happens if we deal with larger objects and distances. When the distance is more than roughly a mile, we rarely walk. Instead, we use tools, such as a bike or a car. Likewise, in science, investigations involving larger distances require more powerful tools. When we observe our own garden, we are happy to see it with an unaided eye (unless one needs glasses); if we observe mountains or a larger landscape, we would gladly have binoculars; and if we want to gaze into the night sky, we employ a telescope (these two exemplify the movement from H toward $s_2$). There are a number of observations that are possible without a tool, but to broaden our knowledge (and to extend the limits of science), we need to employ tools. The macroscale in scientific terms begins when we need a telescope (today usually a radio telescope) to make new observations. Similarly, the microscale in scientific terms begins when we need to employ a microscope to pursue a new discovery.

Axis HK represents the amount of knowledge that science provides. The higher we are on this axis, the more knowledge is accessible to us through the scientific method.[15] We see in Diagram 5 that the further we move away from the human scale, the less is the amount of knowledge available to science. All possible scientific knowledge is therefore limited by curve $s_1-k_1-s_2$ and represented by the gray region. Curve $s_1-k_1-s_2$ is like a horizon that represents the scope of scientific knowledge in terms of space. The use of more and more powerful tools and methods over the centuries has caused a great expansion of the curve.

---

15. Someone could claim, though, that because science deals with particulars, there shouldn't be any limit for scientific discovery in the human scale (i.e., point $k_1$ is impossible), because there can always be another experiment and another conclusion that will add something to science. To this objection we reply that the amount of knowledge refers not to the number of phenomena under scrutiny (this is virtually infinite) but to the scope of knowledge, that is, the amount of physical reality covered by science. Even though science deals with particulars, it is interested in drawing general conclusions and establishing general laws. Therefore, even though the number of particulars with which science could possibly deal is virtually infinite, the number of laws and general conclusions is finite.

## The Microlimits of Science

Now let's see how our diagram corresponds with real scientific discoveries. Vector H→$k_2$ represents knowledge provided by science in a microscale. An example of this can be the 1952 discovery, by Francis Crick and James Watson, of the double-helical structure of DNA. Once the double helix is known, science cannot move further in that same respect (of course, under the condition that the discovery was not mistaken). Obviously, science can address a number of additional issues regarding the structure of DNA. It can find more connections between DNA and other parts of intracellular machinery—but once the true discovery is established, it is done forever. This is why vector H→$k_2$ touches the limit of science, represented by curve $s_1$–$k_1$–$s_2$. Another example is the discovery of the atom. Vector H→$k_3$ is more horizontal because atoms are smaller structures than the molecules that make DNA (i.e., we are moving to a smaller scale). Thus, the discovery of the atom goes closer to $s_1$, but still not touching it, because now science knows of even smaller, subatomic structures. As subsequent models of atoms were presented in the twentieth century, vector H→$k_3$ was approaching curve $s_1$–$k_1$–$s_2$. Although in the case of the atom we do not touch point $s_1$ (an absolute spatial limit of science in microscale), we already see the limit of science because the existence of the atom is not provable in the same way as, for instance, the existence of a living cell. This is why science has been offering different models of the atom, but it cannot determine whether the latest model is the ultimate one or there will be another in the future that will explain data even better. Thus, even the existence of such a relatively large structure (for contemporary physics) as the atom remains a theory and evades a complete scientific explanation.

After discovering the atom, physics went on to discover the structure of the atom, focusing on three subatomic particles—protons, neutrons, and electrons. Later it was proposed that there are smaller particles underlying protons, such as quarks.

Now, the question to ask is: Can we establish a definite point $s_1$ that is an absolute spatial limit of scientific discovery in the microscale? In other words, is there a moment at which science can determine that it cannot go below a given particle? One thing we know for sure is that the lower science descends into the microscale, the greater is the role of the theoretical component in establishing a new discovery. It seems, therefore, that the existence of the ultimate microparticle would be a pure theory, deprived of any possible experimental corroboration. And this is enough to justify point $s_1$, because when experimental science turns into pure speculation, it ceases to be experimental science. It turns into philosophy. Hence, when we look for an ultimate particle—the smallest possible element that underlies physical reality—we need to move from experimental science to philosophy.

Can philosophy establish point $s_1$? Our quest for the smallest particle cannot be infinite, because if we were to find ever-smaller particles, the smallest particle would need to be infinitely small. If the particle were infinitely small, it could not create a body (i.e., a dimensional particle). We know, however, that particles are bodies (in a philosophical sense), so they have dimensions. Therefore, the smallest particle must be finitely small.

We can, however, consider a different solution. Let us suppose that the smallest particles are infinitely small. Perhaps such infinitely small particles could occupy space by means of a mutual interaction rather than by physical dimension. But this solution encounters a problem: The mutual interaction must be due to some energy or force, or any other quality in the supposedly infinitely small particles. Yet, if the smallest (non-dimensional) particle has any quality, such as the force that keeps it in a spatial position relative to other particles, there must be something in the particle that causes it to be like this, and this something has to be different from the particle itself. Otherwise, the particle would be absolutely simple and would not have any quality. On the contrary, if there is something in the particle underlying its quality, the particle cannot be completely simple. If it is not absolutely simple, it is not the smallest possible particle, because there is something more basic in the particle that provides its qualities and makes it complex. Hence, we end up in a paradox: The smallest particle cannot have any qualities and must be absolutely simple. Yet, if the smallest particle is absolutely simple, then it has no qualities, no dimensions, and it cannot generate any force. And if it does not have any dimension or any quality, it is nonexistent. Conclusion: the existence of energy or matter (whichever is more basic for the physical order) is impossible. In our opinion, the only way to overcome this paradox is to postulate a supernatural power that constantly makes possible what is physically impossible, namely, the very existence of matter and energy.

We see here how the physical universe taken in a microscale is inexplicable within its own domain. The limits of science are overcome by a philosophical speculation. Philosophy, in turn, leaves us with a nonmaterial power that we need to postulate in order to justify the very existence of the material universe. This, in turn, implies the necessity of theology.

## The Macrolimits of Science

Ancient and medieval astronomers spoke only about the solar system in somewhat realistic terms. When they described the rest of the universe, they relied more on imagination than observation. Most of them imagined the existence of one or many celestial spheres packed with stars. After Nicolaus Copernicus, there was much discussion about the best model of the solar system, but the realm beyond the solar system was essentially untouched until the eighteenth century, when a German astronomer, William Herschel, discovered the mobility of stars (binary sidereal systems). It was not until the twentieth century that astronomy reached out to the billions of galaxies existing in the universe. The first theories of the universe as a whole (based on physical observations) were developed in the 1920s, especially after Georges Lemaître (and later Edwin Hubble) discovered that galaxies escape from one another and from the earth. This observation led to the Big Bang theory, which gained even better corroboration after the discovery in the 1960s of background microwave radiation. The question is: Are we capable of establishing a definite point $s_2$, i.e., the limit of science regarding the expansion of human knowledge in terms of space in the macroscale?

The basic principles needed to answer this question are the same as those used in establishing $s_1$. We know quite a lot about the so-called "observable universe." It is a universe that we can *in principle* observe. In fact, we currently have not yet observed the objects that are at the edge of the observable universe. It is reasonable to assume that with advancing observation technology, objects that are more distant (and consequently closer to the edge of our observable universe) will be discovered.[16] The limits of the observable universe are established by the distance that light could have traveled since the time when the travel of light became possible. After the Big Bang (about 13.8 billion years ago), for the first roughly 0.3 billion years, matter was too dense to allow radiation to move around freely. Moreover, the calculation needs to account for the expansion of space itself. After taking into account all the data, scientists estimate the

---

16. As of 2016 the most distant object observed was a galaxy classified as GN-z11, about 32 billion light-years away.

radius of the observable universe as 46–47 billion light-years. This is the distance that we can theoretically cover with our observation of the sky in any direction from the earth. Within the observable universe we gather data that can be verified and refined by more exact instruments. The observable universe is the proper object of science.

Now, when we move from research within the observable universe to questions pertaining to the "global universe" (the entirety of the physical universe), speculations abound and take over the discussion. The consensus among scientists on basic statements regarding the observable universe strikingly contrasts with the amount of controversy regarding the global universe. There is not even agreement as to whether the observable universe is identical with the totality of the universe or perhaps the global universe exceeds the observable universe. The latter position is more common, but even then there are still three questions that seriously divide scientists: (1) How large is the global universe if it is larger than the observable portion? (2) Is the global universe finite in terms of space? (3) Are there any other universes outside of ours? We will address each question separately in order to establish the $s_2$ limit of science.

(1) Most scientists agree that the global universe is larger than the observable part of it. It seems that the very idea of an observable universe requires something like an *unobservable* universe, because the limits of the observable universe are not definite.[17] But establishing the size of the global universe goes beyond current data and perhaps beyond the scientific method altogether.

(2) The tentative place to posit the ultimate macroscale limit of science ($s_2$) is somewhere between the answer to the first and the second question (the limits of the observable and the global universe). When our inquiry goes beyond the observable universe, it turns into speculation. This is no surprise, given the basic principle of science which is to

---

17. The limits of the observable universe are established based on the distance to galaxies. Some galaxies are too distant for their light to have reached us within the entire existence of the universe. Some galaxies, though, will become visible in the future, when their radiation reaches the earth. So, the limits of the observable universe should expand over time. However, there is also a contrary process—the more distant a galaxy is, the faster it moves away from us, such that the light from some galaxies will never be visible from the earth. The distance of the galaxies whose light (now emitted) will never reach us is calculated at 16 billion light-years although we can still see the light that these galaxies emitted in the past (so called "event horizon"). Since the very limits of the observable universe are not quite definite, is seems more probable that the totality of the universe is larger than the observable zone. How much bigger escapes science.

build theories based of data. However, beyond the observable universe no empirical data is available. Of course, astronomers would not refrain from extrapolating the cosmological principles governing the observable universe to the entirety of universe. But at this point conclusions are not certain and impossible to test by a fully-fledged scientific method.[18]

The problem of the macroscale limits of science is analogous to the problem of microscale limits: the more distant the reality that is the object of scrutiny, the more speculation in science is involved. When it comes to the ultimate limits of the physical universe, the question must be a purely speculative one, and thus it is philosophical rather than scientific in nature. Hence, to properly address this question, we need to switch to the philosophical mode of reasoning. Otherwise, we stretch the scientific method beyond its scope and fall into naturalism (by expansion).

In order to transit from the scientific astronomical thinking to philosophy of the macroscale, first we need to suspend our view of the universe as proposed by modern science. The scientific view collapses space and time into one reality in order for the equations to work. Astronomy deals with perceptions of the celestial objects, but the problem is that the further away the object, the more detached our observation from its current state. For example, if we look at things which are as distant as the Sun the time for light to travel to us is about 8 minutes and 20 seconds. This means that what we see now happening with the Sun actually happened about eight minutes ago. This is not such a great discrepancy, so it does not bother us. But when we observe the objects millions of light years away, we see them as they were at the moment when they emitted the radiation reaching us only now. Thus, we deal with discrepancies on the order of millions and billions of years. We know from our earthly experience that over a long time many things in nature may substantially change. In fact, humanity did not even exist when most of what we are observing in the sky was actually taking place. Science is not bothered by this huge leap between the actual universe and how it appears to us,

---

18. For example, a popular writer in astrophysics, Ethan R. Siegel, in his article "How Large Is The Entire, Unobservable Universe?," concludes with the following words: "Unless inflation went on for a truly infinite amount of time, or the Universe was born infinitely large, the Universe ought to be finite in extent. The biggest problem of all, though, is that we don't have enough information to definitively answer the question. We only know how to access the information available inside our observable Universe: those 46 billion light-years in all directions. The answer to the biggest of all questions, of whether the Universe is finite or infinite, might be encoded in the Universe itself, but we can't access enough of it to know."

because in science all theories and calculations are based on appearances (empirical data). The appearances of the universe constitute "the universe" for science.

But a simple philosophical question immediately arises: How sure can we be that what we observe in the sky is the actual universe? We know that stars exist only for a given amount of time, usually a few billion years. Thus, if a star emitted its light, let's say, 10 billion years ago, it almost certainly does not even exist now, even though it is a proper object of science. Astronomy is not bothered by this problem, because, as we said, the entire science is based on the observations, that is, the appearances of things as they now present themselves to us. They do not need to correspond with reality. It is enough that the theories match the appearances which constitute the database. For astronomy the universe is younger at larger distances and it does not matter whether it is actually younger or not.

The realistic philosophy, however, can (and should) go beyond the appearances and ask about the *actual* state of the universe; not how it appears, but how it actually is now. Of course, philosophy will not tell us how those cosmic structures currently look like, or whether they even exist. This is not the goal of a philosophical inquiry and it is not possible to know these things outside of the scientific observation. Philosophy is to address a more general problem of the ultimate limits of the global universe: is it finite or infinite? As we said, this question cannot be addressed scientifically, because the scientific method has already ended at the first question concerning the limits of the observable universe. The limits of the unobservable universe, by the very nature of the problem, must be addressed philosophically.

Realistic philosophy is based on our natural experience of the universe by which space and time are two separate categories. Therefore, in a philosophical inquiry (into the ultimate spatial end of the universe) space and time cannot be collapsed into one. Mathematical tools, such as Minkowski space and Einstein's relativity, proved staggeringly successful in describing the phenomena of the observable universe at large distances. But there is no reason to assume that when asking about the actual universe (as it is now, not as it appears now) relativity would work as well. On the contrary, it seems that relativity and the time-space fold work well (as cosmological models) precisely because we have access only to appearances that were generated long ago. In a way, the fact that with growing distance we observe further past gives us a glimpse into the history of the universe but at the cost of knowing how it is now. It is reasonable to assume that the more distant the object the greater the discrepancy between its appearance to us and its actual current state (because in the cosmos everything constantly change). Accordingly, when we abandon appearances for the sake of reality, we go back to understanding the universe as a kind of bulb of space, as it had been postulated in a stationary model. In the current astronomical categories, the bulb would be growing in diameter as the universe expands, but we need to imagine it in a frozen state, as if it were not expanding. This leaves us with a common-sense model of the universe akin to the one imagined by Aristotle (and generations of other scholars from antiquity to the early twentieth century). Paradoxically, this model, which can be approached only philosophically, is closer to what the real universe *is* like (not *looks* like).

Now, the finitude of the universe can be considered either as a finitude of the volume or space. By spatial finitude we understand the idea that if a material object could be instantly repositioned to the most distant place from where it is then the object would be moved finitely far away. But if space were infinite then the object would be repositioned infinitely far away.

Most astronomers would tell us that the question about the spatial limits of the observable universe makes little sense, because the universe is "finite but boundless." We could infinitely travel in one direction and never reach the end. At the same time, astronomers would say that the observable universe is finite in terms of volume (which is currently estimated at $3.57 \times 10^{80}$ m$^3$). To imagine this, astronomers invite us to think about the universe as a sphere that is growing, like a balloon that is being pumped. An object in space behaves as if it was placed on the surface of the balloon. The universe expands but it is finite in volume. At the same time the object on the surface can travel infinitely in whichever direction. We need to remember, however, that this popular picture is just a model, a metaphor of how the observable universe can be imagined based on the appearances. In philosophy we do not stop at appearances and metaphors; we try to grasp the reality itself.

At the philosophical level of inquiry, we can inquire into the spatial finitude of the universe, because space is neither curved nor is it mingled with time. And if we could instantly move (not travel at the speed of light, but instantly be repositioned) to the place in the universe most distant from us, then either we could move infinitely far from here or we would end up next to a kind of a border that would constitute an ultimate end of the universe. In reality, there is no other option. But this reality cannot be addressed by strict natural science. Ultimately no data is and most probably never will be attainable regarding the finitude of the global universe. The question of the spatial limits of the universe reveals the limits of the scientific method as such.

(3) The third question—whether there are other universes besides ours—cannot be properly addressed by science either. The multiverse hypothesis cannot be scientific, even if it is proposed and promoted by scientists of the highest stature. They simply exceed, either consciously or not, the domain of science and begin philosophizing. Even though science cannot establish the precise position of $s_2$, it can establish the tentative range of scientific answers regarding the limits of the observable universe. A more speculative science, such as cosmology, may perhaps theorize about the global universe. But the third question is outside of the range of even most speculative science. Claims such as "data force us to assume multitude or an infinite number of universes" are misleading and false. There are two reasons why it is so.

The first is implied by a commonsense definition of science stating that science is a discipline that deals with the physical universe. Any claim going outside of the physical universe is beyond science and, in this sense, unscientific. There is, however, a logical trick here: The proponents of the multiverse sometimes say that science deals with the entirety of the universe, which consists of many or an infinite number of universes. Then, the multiverse would be the proper object of science. But in this kind of reasoning the conclusion resides in the premises and renders the argument circular. A proponent of the multiverse assumes in the first place that the universe is the multiverse, but this is what should be proved, not assumed.

The second reason why the multiverse hypothesis is not scientific is that it is entirely untestable, which means it can be neither verified nor falsified by means of the scientific method. Scientific method theoretically extends to the entire physical universe, but practically it is limited to the observable (and by this testable) part of it. So, if a hypothesis remains within the observable universe, or can be anyhow tested by the experimental method, it can be counted as hard science. If it goes beyond the observable universe, but remains within the universe, it can belong to a speculative type of science that already entails some elements of philosophy. But if a hypothesis goes beyond the universe it should not be counted as scientific in any sense of natural science.

## Philosophy of the Macroscale

Based on our previous considerations, we can see that philosophy is a more appropriate discipline for speaking about the global universe and explaining whether it is spatially finite or infinite. Therefore, having established the tentative limits of science, we are now switching to philosophy.

In order to inquire into the spatial limits of the global universe, first we need to define space. We know from both daily experience and advanced science that space is a combination of particles and a void. Usually, material particles are very small compared to the amount of void in which they are dispersed. For instance, the nucleus of an atom is very far from the electrons, and the electrons are infinitely small compared to the distances they travel when moving within the region surrounding the nucleus. Thus, the atom, with respect to its volume, is mostly a void. The same basic structure is represented in the solar system, galaxies, and the universe as a whole.

Do we need both a void and particles to create space? Perhaps at some stage of the expansion of the universe, particles were packed so densely in space that there was no void between them. Hence, we could possibly assume that there can be space created by particles alone, i.e., particles without a void. In contrast, we do not have any experience of space without particles. Even in the great voids of the universe, there is radiation and there are particles. Therefore, there are no real voids, nothing like truly empty spaces, in the universe. Space must always be "occupied" in the sense that a void must be somehow affected by particles, in the form of either radiation or physical bodies, or at least physical forces such as gravity. Thus, according to our best knowledge, space cannot exist (in reality) without particles. On these grounds, we can define space as a void combined with particles which is measurable by physical distance. The question about the limits of the universe boils down to the problem of the finitude of space: Is space finite or infinite?

Many philosophers over the centuries have asked this question. It seems that classically minded scholars reject the idea of infinite space.[19]

---

19. Thomas Aquinas provides a fully-fledged answer to the question of whether a physical body can be infinite and of whether there may actually exist an infinite number of things. In what follows, I am inspired by his argumentation, but I have tried to reduce it to one narrow path that is the most relevant to our discussion. For his other arguments, see *Summa Theologiae* (hereafter *S.Th.*) I,7,3–4. Cf. also Aristotle, *Physics*, 3.6.

The reason is that space is something that has dimensions, and something having dimensions cannot actually be infinite.

To better understand this, we can use an example derived from mathematics. In calculations, we commonly use a symbol representing infinity, i.e., ∞. But is the symbol itself infinite? Surely it is not. It is as tiny and limited as many other symbols used in a written language. This symbol refers us to an idea that we have in our minds—an idea of infinity. In fact, we cannot even imagine infinity; we imagine perhaps some immense cloud or ocean, but such things are finite. Another method of presenting infinity in mathematics (besides using the symbol) would be to enumerate all natural numbers from one to infinity. This, however, can never be accomplished, because to reach infinity we always need to add another number (n+1) after the last number (n) in the enumeration. Representing infinity in this way would take an infinite amount of time and could never be accomplished. In this sense, it is impossible. This is why we use the first method; instead of enumeration, we use a symbol that refers to something in our minds, even if this something cannot be imagined. We can conceive of infinity only by abstraction from everything corporeal we know.

It is similar with space. An infinite space would require infinite distances. But to have infinite distances in reality (not just in the mind), an infinite number of measurable (finite) distances would need to exist in reality. To exist in reality means either to be created from nothing or to transit from potentiality to reality. In either case, an infinite number would need to be somehow realized, and this is impossible in the same way as is mathematical enumeration up to infinity. This is why the really existing space must be actually limited. However, this finitude does not exclude the possibility that space could grow infinitely over an infinite length of time, if such time were available.[20] Space, therefore, can be infinite only potentially; actually, it needs to be finite.

---

20. For centuries, philosophers offered many arguments for the impossibility of an infinite space. One way to demonstrate it is by means of geometry. If we actually had an unlimited space at hand, we could draw a triangle whose vertices would be infinitely distant. We could slice the triangle with lines parallel to the base, for example, one meter from each other. Then we could measure the length of the consecutive lines beginning from the base up to the top vertex and then from the top vertex down to the base, until we meet in the middle. If we measured the length of the parallel lines from the base up to the top vertex, they would all be infinite, because the base is infinite and the first line parallel to the base is just a little bit shorter. But if we start measuring the length of the lines beginning from the opposite side (i.e., from the top vertex down to

In contemporary discussions about the boundaries of the universe, the arguments of philosophers do not play an important role. The question has been relegated to the speculations of astrophysicists. However, when it comes to the global universe, science collapses due to the lack of data. In fact, even within the observable universe, when addressing the problems of large distances, astronomy resorts to mathematical models and analogies that depart from our common experience and oftentimes do not settle the questions. The further distant the objects studied, the more speculation is involved in cosmological studies. In contrast, philosophy, by resorting to the common-sense model, enables us to establish that the universe must be spatially finite. If this is the case, there is another question that naturally follows, namely, what is beyond our universe?

Since the universe is finite, there are just two possible options: either there are other universes, or there is nothing at all outside of our universe. The first option falls into the same problems as the idea of an infinite universe. If there is another universe, we must ask if there is still another universe outside of that one, and so on up to infinity. Philosophy allows for multiple universes, but not for an infinite number of them. Since we can learn only about our universe and have no access to any imagined universes beyond ours, it would be superfluous to assume their existence. We cannot even demonstrate whether or not they exist, let alone say anything about their properties. So their existence is—at best—irrelevant for us.

Whether there are many universes or just one, in either case we face one serious problem: since the totality of universes is finite, there must be a limit of space, a "place" where space simply finishes. In other words, it is a place where something touches nothing. Since space has dimensions and nothingness has no dimensions, they cannot be neighbors. Adding some intermediary bodies does not help, because ultimately, whatever the body might be, it must have dimensions and must touch nothingness that does not have any dimension or any other properties. Nothing can border with nothingness. In our opinion, the very existence of a finite

---

the base), they would all be finite. The first would be quite short, and each next line would grow in length only a little bit. Eventually we would come to the point where a finite line is adjacent to an infinite one. This, however, is absurd, if we understand what a triangle is. This thought experiment demonstrates that an infinite plane leads to logical contradictions. The same applies to an infinite space; therefore, an infinite space is impossible. By the way, for Thomas Aquinas, one of the arguments affirming that the universe must be finite stems from the belief that the universe was created according to a clear and definite idea in God's mind. And infinity is not this kind of idea.

universe is therefore impossible under natural conditions. But an infinite universe is not possible either. Hence, we need to postulate the existence of some immaterial supernatural power that constantly makes it possible for the universe to neighbor nothingness, i.e., to be spatially finite. Since the distance between nothingness and something is infinite, the power that creates the limits of the universe must also be infinite in order to overcome the distance.

Here is the limit of philosophy. If we glance toward theology, we can say that the universe as a whole needs to exist in God in order to be finite and thus be something for which existence is possible. The old truth that the universe was born in the divine mind and is maintained in existence by the constant action of God now gains an additional meaning: the universe needs to be "in God", i.e., needs to be constantly supernaturally confined by God in order to be physically possible.

*The Spatial Limits of Science—A Summary*

According to Diagram 5, science has limits in both directions—micro- and macroscale. The further it strays from the human scale, the more the empirical component is replaced by the theoretical component, up to the point at which science becomes pure, untestable speculation. This is the point of an ultimate limit of science ($s_1$ for the microscale or $s_2$ for the macroscale).

One can, however, raise an objection against Diagram 5: Why not allow an unlimited growth of science? Back in the sixteenth century and earlier, many scholars believed that it was not possible to reach beyond the visible sphere of the fixed stars. Before the microscope was invented, people thought it was impossible to observe anything smaller than what the best human eye can see. How do we know that science has limits at all? Based on our historical experience of the never-ending progress in science, shouldn't we rather extrapolate from this experience and assume that there is nothing like points $s_1$ and $s_2$? In other words, shouldn't we assume that along with the increasingly better observational tools and progressing theories in various disciplines, science will expand and cover ever-new fields of inquiry?

To answer this question, we need to realize that science can progress in three senses: the first is the progress of knowledge about the objects that we already know, the second is discovering new objects, and the third is the progress of knowledge regarding what may be known in scientific terms. At the early stages of scientific development, knowledge was growing only in the first sense—scientists (philosophers of nature) were discovering properties and laws governing the objects of daily experience and those that could be observed with the naked eye. Later (especially beginning with early modernity and the first scientific revolution), science developed into the discovery of new objects (in both microscale, like the discovery of cells and later atoms, and macroscale, such as the discovery of new planets and galaxies), and only relatively recently (in the twentieth century) did science discover that there are insurmountable limits of science. Hence, our claim about the limits of science would not have been valid in the sixteenth or seventeenth century, when there was a great deal of the first two types of growth but not the third. In contrast, today we can speak about the ultimate limits of science, because the scientific method itself shows us the limits of progress in the first and second sense. (The same objection and response apply to Diagrams 6, 7A, and 7B, which we will present next.)

b. The Limits of Science in Terms of Time

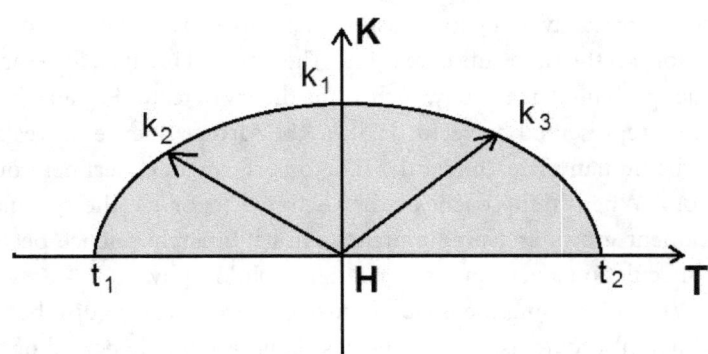

**Diagram 6. Scientific knowledge with regard to time (diachronous approach).**

In Diagram 6, we see the tentative limits of science with respect to short and long periods of time. Analogously to Diagram 5, we have the human scale, which accounts for roughly fifty to a hundred years, near point H. This is the length of time that we normally experience during our lives. The more we move to the left on axis HT, the shorter are the periods of time being considered. When we move to the right, we move from hundreds to thousands to millions and billions of years. The periods of time on both sides can refer to either the past or the future; we consider shorter periods of time into the past or the future when we are closer to $t_1$ and longer periods into the past or the future near point $t_2$. Diagram 6 represents the *diachronous approach*, which means that we look at events (of short or long duration) at theoretically any time in the history of the universe. Admittedly, science can provide the most certain (i.e., tested and verified) knowledge about events whose duration is within the human scale. The expansion of science in this dimension is represented by vector H→$k_1$. Theories falling into this category are Newtonian mechanics, Einstein's relativity, and all other theories applicable now and in the human timescale. The progress of science in this context would be depicted by the extension of vector H→$k_1$. The longer the vector, the more explanatory power a theory has. For example, Newton's mechanics would correspond to a shorter vector compared to Einstein's relativity. Vector H→$k_2$ symbolizes theories, such as those regarding electromagnetism and radiation, that explain phenomena taking place in a duration shorter than the human scale, but still not reaching the shortest possible periods of time. Even closer to $t_1$ are theories from quantum mechanics which deal with phenomena happening usually in fractions of a second. Analogously, theories involving long-lasting phenomena would be represented by vectors on the right side of axis HK. These would include, for example, theories presented in geology, such as the theory of tectonic plates (which could be represented by vector H→$k_3$). The further science moves away from the human scale, the harder it is to perform an experiment or test a theory. When it approaches either extreme ($t_1$ or $t_2$), the speculative component grows and predominates until ultimately science becomes pure speculation and requires transition to philosophy.

When discussing the issue of time, we need to distinguish between five different problems: (1) the shortest/longest possible period of time; (2) the shortest/longest measurable period of time; (3) the shortest period of time we have actually measured; (4) the shortest possible event duration; and (5) the actual duration of an event. Diagram 6 cannot be applied to the first problem, because most probably there can always be a shorter period of time in the sense that any given fraction of a moment can always be divided into two shorter moments. Also, the longest period of time can always be longer because the duration of the universe could continue indefinitely into the future. Hence, the diagram does not describe the first problem—if it did, then points $t_1$ and $t_2$ should not be there at all. The fifth problem concerns the actual duration of an event that is an object of science. The actual duration of the event can differ from the scientific measurement thereof. The goal of science is to provide the most accurate measurement, and the closer the measure is to the actual duration, the more true is the science. (This is the problem of truth in science represented by $t_5$ in Diagram 1.) Once the duration of an event is measured, science uses the measured time, not the actual duration of the event. If the measurement is perfectly accurate, then both times are equal. Sometimes science can know with some degree of certainty the possible inaccuracy of the measurement. Then, the corresponding range of numbers will be taken into account in calculations and theories. In some cases, though, the measurement is satisfactorily accurate, even if it is not exact. We see, therefore, that in some cases science does not deal with the actual duration of an event; in other cases, it does not have to deal with the actual time; in still other cases, it is impossible for science to measure the actual time. Thus, the fifth problem is not accounted for in Diagram 6. In contrast, problems (2), (3) and (4) are included in the diagram, though each of them may have different values for $t_1$ and $t_2$.

Currently, it is commonly accepted that the shortest time that has any physical meaning is the Planck time, roughly $5.4 \times 10^{-44}$ seconds. Most scientists would say that this is the shortest possible time in the second and fourth sense. Therefore, this number stands for $t_1$. As of 2010, the shortest measured time was on the order of 12 attoseconds ($1.2 \times 10^{-17}$ seconds), about $2.2 \times 10^{26}$ Planck times.[21] Since this is a very long time compared to one Planck time, we can expect that the measurements of the future will slowly come closer to one Planck, most probably never reaching it. In any case, if we consider Diagram 6 with respect to the third problem, the current $t_1$ limit is about 12 attoseconds.

In order to properly present $t_2$, we need to switch from the *diachronous* mode to the *synchronous* mode of presentation. This is because $t_2$ differs depending on whether we look into the past or into the future. Diagram 7A represents the synchronous approach.

---

21. Cf. http://phys.org/news/2010-15-attoseconds-world-shortest.html (9 August 2015).

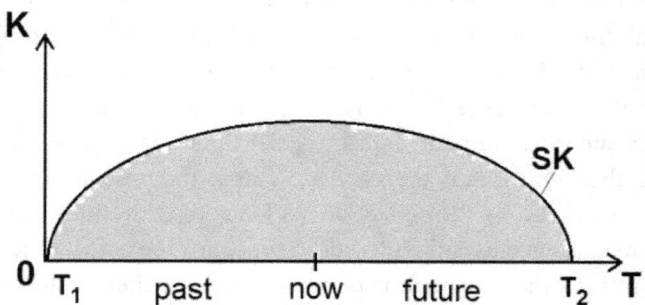

**Diagram 7A. Scientific knowledge with regard to time (synchronous approach).**

Axis o→T represents the timeline from the Big Bang through the present into the unknown future. The gray region under curve SK represents the scope of attainable scientific knowledge. For the sake of distinction between quantities in Diagram 6 and Diagram 7, we now use capital letters ($T_1$ and $T_2$) for $t_1$ and $t_2$.

Are there any limits of science in terms of addressing the past and the future? If there are, they should be limited to the duration of the universe, because, as we said, science cannot address problems falling outside of the physical universe, regardless of whether we consider the universe in a spatial or a temporal sense. Based on our knowledge about the universe, we estimate its age at about 13.8 billion years. There is, however, an inherent limitation in our understanding of the past, and this limitation is based on the Planck time. If one Planck is the shortest time relevant to physical events, then we can never investigate the universe before it was one Planck old. The moment when the universe was one Planck old is called the initial singularity, and it delineates the limits of science. This physical restriction also makes logical sense, because if we compare the beginning of the universe to a great explosion, there must be something to explode. Since nothingness cannot explode, the Big Bang must begin with something. This was that tiny, one-Planck-old cube of space of extremely high, yet finite, energy and density. Whatever scientists say about the events preceding the initial singularity is pure speculation that is untestable and outside of any empirical corroboration. For this reason, it cannot properly be called science, even if the authors of these speculations are scientists who are inspired by their scientific knowledge and perhaps use scientific language.

The initial singularity marks point $T_1$, which delineates an ultimate limit of science. If we want to expand our knowledge beyond $T_1$, we need to resort to other forms of knowledge, such as philosophy and theology. For the sake of coherence with our theological knowledge, it seems reasonable to assume that if we were to move back in time before the initial singularity, the tiny cube of matter of extreme density would be even smaller and denser, and before that there was nothing that turned into something. These are, however, not scientific but rather philosophical conclusions. The idea of the Big Bang originating from nothingness perfectly harmonizes with the Christian theological idea of *creatio ex nihilo*. Christians should not be surprised that the two books (see I,1,c) supplement each other with such a smooth transition and such clear logic. Indeed, this is what we would expect if Christian revelation were true.

Next, we consider the future. We know that theoretically time could progress infinitely. This does not mean that time would be actually infinite (impossible for reasons analogous to those regarding the impossibility of an infinite space; see the previous section). It means only that time could pass forever along with the existence of the universe. Physical data strongly favor the so-called "heat-death" of the universe, but they do not tell us that after the heat-death the universe would cease to exist. Based on the current processes observed in the universe, we can say that the total entropy is increasing, which means that organized structures such as atoms, planets, and galaxies decay and slowly disappear. By extrapolation, we can predict that in some very distant future all stars will have burned out. Even black holes will also have fallen apart (as a result of radiation) into a dilute gas of photons and leptons. In this state, there will be no organized structure in the universe whatsoever; the cosmos will have turned into "perfect" chaos. Scientists estimate that the universe will reach heat-death in about $10^{100}$ years. This is, therefore, point $T_2$ in Diagram 7A.

Establishing the exact number of years in this case is not essential to our argument. The bottom line is that any scientific predictions going beyond the heat death of the universe are untestable and purely speculative. Naturalism occurs when a scientist claims that science supports the idea that the "dead universe"—chaos—will return to any form of cosmos thanks to natural processes known in the current form of the universe. This kind of claim, even if presented by a scientist, with the use of scientific language and in the name of science, cannot be considered scientific. Properly speaking, this kind of prediction is a matter of philosophy or an expression of a personal belief. In the latter case, we can say that it is a theological claim. Naturalism begins when the scientific method is extended beyond one of the four absolute limits of science ($s_1$, $s_2$, $T_1$, $T_2$). If this happens, science falls into naturalism by expansion (according to our definition in I,2,a).

Yet, this is not the only way in which scientists can fall into naturalism. Naturalism occurs whenever the scientific method is extended beyond the region of scientific validity, marked by curve SK. In what follows (I,4,c), we will show how the scientific method may be extended beyond the scope of science (thus falling into naturalism) without violating any of the four absolute limits of science.

### c. The Limits of Science and Biological Macroevolution

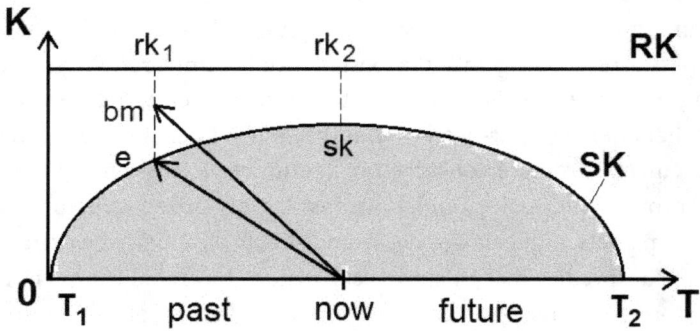

Diagram 7B. Scientific knowledge and religious knowledge (synchronous approach).

Diagram 7B is just a developed form of Diagram 7A. The white region under line RK represents religious (and to some degree philosophical) knowledge about the universe ($t_2$ in Diagram 1). The gray region under curve SK represents scientific knowledge, and thus the curve itself represents the expected compatibility between and a delineation of the two types of knowledge derived from the two books (i.e., nature and the supernatural revelation). We see that the further into the past or future we look, the less we can know based on science alone and the greater are the philosophical and theological components of human knowledge (i.e., the distance between curve SK and line RK increases). The extreme points $T_1$ and $T_2$ are particularly clear: we cannot reach scientifically before the singular point, but theology tells us that God created the universe out of nothing, thereby establishing the beginning of space, energy/matter, and time. At $T_1$ our entire knowledge comes from theology. When we look into the future, we see that science can predict with greater probability events closer to us in terms of time. For instance, all science—indeed, all human activity—is based on the prediction that tomorrow the sun will rise again and that physical laws will remain unchanged. Astronomy can with great precision predict some celestial events (such as solar eclipses) thousands of years into the future. Yet, these kinds of predictions cannot address any time in whatever future. We know that owing to the expansion of the universe the distances between planets change, thus making infinite astronomical predictions hard or impossible. Also, owing to ever-increasing entropy, all laws of nature must ultimately collapse, such that predictions regarding physical phenomena in the distant future are unreliable or even impossible. We also cannot dismiss the possibility that some other, currently unpredictable events, will enter the scene and alter the currently predicted sequence of causes and effects. The further science wants to see into the future, the more speculation is involved. Thus, even before reaching $T_2$, there is a growing component of philosophical speculation regarding the future of the physical universe. We see that curve SK gradually approaches a knowledge level of zero. Based on science alone, we know that the universe is destined to complete destruction and then, most probably, infinite existence in the form of "perfect" chaos, where no structure, no life, no chemistry, and no specified form can exist. This seems to be quite a pessimistic fate. Theology, however, tells us that God, after finishing human history, will re-create the universe and make it everlasting. At both endpoints $T_1$ and $T_2$, theology provides more knowledge than science and philosophy combined.

Someone could ask, though, why curve SK does not touch line RK. It seems that science is potentially able to explain all of the physical universe around point "now," i.e., the current state of the universe or human history. Instead, we see distance sk–$rk_2$. The reason for this is that there is a type of knowledge coming from theology that explains some phenomena happening now in the visible order but surpassing the physical order, such as miracles, supernatural prophecies and visions, and the like.

An attempt to extend the curve up to point $rk_2$ would constitute second-grade naturalism (see I,2,b).

In Diagram 7B, vector now→e represents the theory of biological evolution. This is a theory proposed by scientists to explain the origin of different species of living beings. As such, this theory is historical in nature. It does not belong to "now," because it cannot be observed or tested within the period delimited by human history, and certainly not within the human scale. If evolution is understood as a theory that attempts to account for the origin of all biodiversity, it should be called biological macroevolution.[22] The history of life on earth, as we know from fossils, has lasted for 3–4 billion years, with a few moments of dramatic increase in biological forms, such as the Cambrian explosion roughly 540 million years ago. All of the supposed macroevolutionary changes must have happened in a time vastly exceeding the time available to scientific observations and experiments. Scientists try to confirm biological macroevolution by extrapolating what we can see and test in laboratories today. Some scientists would firmly claim, though, that the fossil record directly proves evolution. This is a controversial topic that will be addressed later in the book. The point of interest for us now is the question of whether biological macroevolution remains within the scope of scientific knowledge (the gray region) or violates the limits of science. In other words, is biological macroevolution more properly represented by vector now→e or vector now→bm?

To answer this question, we first need to make our notion of biological macroevolution more precise. Since atheists and materialists do not recognize any true knowledge beyond curve SK, for them no theory can go beyond that curve. For this reason, there is no sense in bringing materialistic or atheistic evolution into our discussion. The only relevant understanding of biological macroevolution in our context is the theistic type of biological macroevolution.

Theistic evolution accepts the general biological postulates about universal common ancestry and the transformation of species, but it also recognizes the existence of God and his participation in evolution. The most common understanding is that God began the cosmic process of evolution, setting up all the properties of the physical universe in such a way as to make evolution possible. Still, he governs this process by his providence and acts upon creation as the final cause of all physical and

---

22. For the definitions of "evolution," "species," and "theistic evolution" that we adopt here, see II,1,a–c and III,1,b.

biological phenomena. Theistic evolutionists say that even though science considers the mechanism of evolution to be completely random, from the theological perspective it is planned and guided by God. Some theistic evolutionists invoke the Thomistic idea of God using chance events to accomplish His preplanned ends to explain how the random nature of evolutionary process fits in a broader theological worldview. We see that in theistic evolution both theological and scientific knowledge is taken into account—part of our knowledge about the origin of species is derived from natural science (everything below point e) and another part from theology (segment e–$rk_1$). Apparently, this is exactly what a Christian approach would require—a synthesis of science and faith.

However, there is a problem with the synthesis adopted in theistic evolution. Scientific theory in our diagram ends with the limits of science at point e. If theistic evolution has not fallen into naturalism, it must be represented by vector now→e, because this is the furthest that science (i.e., natural explanations) can go without falling into naturalism. The fact that theistic evolution accepts *some* religious truths about the universe does not, by itself, ensure that it respects the proper limits of science. To determine if it does, we need to ask: Which particular theological truths does theistic evolution accept, and which of them does it reject?

Theistic evolutionists recognize theological truths such as *creatio ex nihilo*, God's providence over the universe, and God's mediated causality. These theological truths are represented by segment e–$rk_1$. Now, the problem is that all of these truths are also accepted at point now, which stands for the history of salvation. According to Christian theology, God extends his providence over the universe and functions as the first and final cause for all events throughout all of human history. In other words, the amount of theological truth ($t_2$ in Diagram 1) accepted in theistic evolution equals the amount of truth accepted at point now. This means that for theistic evolution, segment e–$rk_1$ equals segment sk–$rk_2$, which should not be the case if theistic evolution remained within the grey region (the scope of science). Theistic evolution, since it is a historical theory addressing events dating to long before now, by its very nature should have a greater theological factor than the theories referring to now. This is confirmed by our diagram, where segment e–$rk_1$ is longer than segment sk–$rk_2$.

Actually, the theological component (distance between SK and RK) in theistic evolution maybe even smaller than the amount of theology occurring at point now, because at now (which stands for the history of salvation) Christianity speaks about the supernatural works of God in the natural order, such as miracles, whereas theistic evolution excludes any supernatural divine activity in the history of the formation of the universe. This means that theistic evolution is not properly represented by vector now→e. To diminish the theological factor in theistic evolution (i.e., shorten segment e–$rk_1$), we need to move the theory "up," which means that the vector representing it must be longer. This is properly depicted by vector now→bm. The distance from bm to RK either equals or is even shorter than the segment sk–$rk_2$. And this adequately represents the concept of theistic evolution. Yet, vector now→bm goes beyond curve SK. It follows that theistic evolution extends the scientific method beyond its scope and falls into naturalism (specifically first-grade naturalism). Later in the book, we will discuss with greater precision which theological truths about the universe ($t_2$ in Diagram 1) are rejected in theistic evolution (see III,3).

## 5. Galileo and Darwin

### a. The Galileo Affair: A Summary

Numerous books and papers have been published on Galileo Galilei. Our goal here is not to contribute anything new to historical study of the Galileo affair. We will address a different problem, namely, the problem of using Galileo's precedent to promote Darwin's theory in Catholic theology. A popular claim among the proponents of theistic evolution is that as science was proved right and theology wrong with regard to heliocentrism, in the same way theology has to give up on the creation of species and accept natural evolutionary explanations instead. Theistic evolutionists deem the acceptance of the evolutionary origin of species a historical necessity, something that logically follows from the development of science. To see if they are right, first we need to understand the nature of the Galileo affair.

The case of the Italian astronomer is multifaceted. For our purpose, we will identify three major issues. The first is the problem of Galileo's scientific activity and attempts to develop astronomy and other sciences. The second is the question of his theology, specifically, how to resolve the biblical argument favoring the stability of the earth in light of new theory favoring earth's movement. The third issue is his personal story—his difficulties with Church authorities and his ecclesiastical condemnation.

The third issue is of the least importance for us, though we will start by making a few comments on it.

We know that much of Galileo's problem with the clergy and the Church hierarchy was generated by his bold claims presenting theories as facts and his disregard for his opponents, to the point of ridiculing them. It is quite probable that the final condemnation of 1633 was a direct consequence of his publication of the *Dialogue Concerning the Two Chief World Systems* (1632). In that book, Galileo presents the arguments for the Copernican system in the form of a conversation between three people. The traditional Aristotelian worldview is defended by a character named "Simplicio," which implies a man of simple mind, or, to put it bluntly, a fool. The problem is that Simplicio effectively stands for the most conservative clergy of the time, including Pope Urban VIII. The book was the straw that broke the camel's back. Aristotelian philosophers, proponents of the geocentric model, constituted the core of Italian academia. Apparently, most of them were not against the Copernican system provided it was proposed as a hypothesis and a mathematical model for facilitating astronomical calculations, rather than a physical reality. Copernicus published his book over seventy years before the first admonition against Galileo was declared. Throughout those seventy years, there was little controversy over heliocentrism and no relevant documents were issued by the Church. Indeed, Copernicus's discoveries were used in the calendar reform carried out under Pope Gregory XIII in 1582. Copernicus was the one who revolutionized astronomy, but he did not fall under any ecclesiastical condemnation; this suggests that it was something other than heliocentrism itself that brought the condemnation on Galileo. The way he acted was not welcoming to opponents. Moreover, the spirit of the time encouraged harsh judgments. It was the time of the Counter-Reformation, not long after the Council of Trent, when the Catholic Church was recovering after the dogmatic and institutional confusion caused by the Protestant Reformation. Church officials of the time were quite fond of Catholic doctrines and convictions. This attitude allowed them to pass strong judgments without much hesitation.

Nevertheless, whatever circumstances accompanied the Galileo affair, the fact is that Galileo was condemned for claiming that the earth revolves while the sun sits still in the middle of the universe. Today we know that neither of these claims is completely accurate, yet, both are closer to the truth than the opposite propositions defended by Church officials in Galileo's time. And this is why Galileo was *more right* than his opponents. Pope John Paul II generously initiated extensive studies to yield an objective and just judgment regarding the famous Italian astronomer. Those studies ended in 1992, after twelve years, with the recognition of the "subjective error in assessment" on the part of the Holy Office. Since then, the Church considers Galileo's case closed, at least as far as the ecclesiastical judgment goes. Contrary to what popular media announced, there has never been any official "rehabilitation" of Galileo. The decrees forbidding Copernican theory were effectively invalidated many years earlier: in 1757, Pope Benedict XIV removed from the papal Index books affirming the movement of the earth, and in 1820 a book by Giuseppe Settele presenting the Copernican system as fact rather than theory received the Roman imprimatur. These decisions overrode the previous condemnations long before the assessment was issued by John Paul II's commission.

When it comes to the first problem—Galileo's science—it is obvious today that many of his postulates were inaccurate or mistaken. His theory of tides turned out to be entirely false, his theory of motion was soon replaced by Newton's mechanics, and even his two major astronomical claims were inaccurate. In fact, neither the earth revolves around the sun nor the sun around the earth. Rather, both bodies revolve around the common center of gravity (more accurately, the center of gravity of the solar system). Nevertheless, to maintain that for these reasons Galileo was not right at all, or that he was rightly condemned, would not do justice to him. We cannot evaluate his physical theories based on current scientific knowledge (this would be the error known in historical sciences as *anachronism*). Scientific theories, by their very nature, are just approximations whose task is to explain a collection of data in a coherent way. If further observation, improved experiments, and additional discoveries make it possible to demonstrate that a given theory is true beyond reasonable doubt, then it gains a status of a "fact." Such facts today (which were not obvious centuries ago) include the globosity of the earth, the existence of eight planets in the solar system, the existence of galaxies and bacteria, the structure of proteins, vitamins, and enzymes, etc. Even if the motion of the earth was not demonstrated by Galileo to the point that it could be called a "fact," he nonetheless presented strong and convincing arguments in favor of the Copernican system over the geocentric one. In contrast to Galileo's scientific attitude, some of his adversaries refused to even look through the telescope. The essence of Galileo's case is not that he was right according to our current knowledge (because he wasn't), but that he was closer to the truth than his opponents at the time. And this fact owes him commendation rather than condemnation.

Our particular interest concerns the second issue that contributed to Galileo's case, that is, the problem of theological knowledge derived from the Bible regarding the physical universe ($t_2$ in Diagram 1). What is the authority of the Bible in establishing physical facts? Does the Bible have any authority over physical facts, and how can we reconcile its teachings with what we know from science? The first decree against Galileo, issued by the Roman Inquisition in 1616, reads:

> 1. The sun is the center of the world, and entirely immobile of local motion.
> *Appraisal*: All have said the stated proposition to be foolish and absurd in Philosophy; and formally heretical, since it expressly contradicts the sense of sacred scripture in many places, according to the quality of the words, and according to the common exposition, and understanding, of the Holy Fathers and the learned Theologians.
> 2. The earth is not the center of the world, and not immobile, but moves according to the whole of itself, and also by diurnal motion.
> *Appraisal*: All have said, this proposition to receive the same appraisal in Philosophy; and regarding Theological truth, at least to be erroneous in faith.[23]

We see that the Inquisition condemned two astronomical claims by Galileo as contrary to both philosophy (which included natural science at the time) and the Bible. His claims were judged to be false according to all three levels of knowledge ($t_2$, $t_3$, and $t_4$ in Diagram 1). Therefore, saying that the cardinals who signed the decree attacked science using theology oversimplifies the case. In fact, they defended the well-established science-faith synthesis of their time against the new, in their opinion wrong, scientific claim. The problem thus consisted of two planes—scientific and theological. Of these two, the more important one for the contemporary dialogue between science and religion is the theological plane. How is it possible that the Inquisition established that Galileo's position was contrary to the Bible, i.e., theologically wrong? The Inquisition established

---

23. Translation based on the original in C. M. Graney, *The Inquisition's Semicolon: Punctuation, Translation, and Science in the 1616 Condemnation of the Copernican System*, but slightly adjusted in accord with the translation of Giorgio De Santillana, *The Crime of Galileo*, 121.

that the Copernican system was formally heretical.[24] Yet, within two centuries, both the science and theology defended by the Inquisition proved to be wrong. Does this not invalidate any authority that the Bible may have regarding physical matters? How can we know that the Church has any authority when speaking about nature?

Precisely because of the Galileo affair, today many theologians believe that the Bible cannot in any way be used to judge scientific theories. According to this perspective, $t_2$ in Diagram 1 is either nonexistent or impossible to establish, or at least it should always be judged by scientists rather than theologians. Moreover, the Church herself is very cautious about judging science or challenging anything affirmed by the scientific community. We can even notice a kind of psychological complex—if not fear—on the Church's part when it comes to the dialogue with science. This complex is deepened every time science triumphs in the modern world of technology; meanwhile, religion is removed further and further away from mainstream culture.

The Galileo affair has been used by some scholars to defend and propagate the theory of biological macroevolution. In their opinion, Galileo's case has given the Church a lesson that should never be forgotten.[25] They lament the fact that even today there are still some believers who oppose the "science of evolution" based on their wrong understanding of the Bible, in the same way as the "science of the planets" was opposed by ecclesiastical scholars in the time of Galileo. According to them, scholars denying evolution are mistaken about the Genesis account of creation in the same way as most theologians were mistaken about geocentrism four hundred years ago. The conclusion of these authors is that the Church should accept evolution and not risk another false condemnation of the "scientific truth." Challenging Darwin, they say, would cause another Galileo affair and permanently discredit the Church in the eyes of modern society. Quite often, these authors also highlight the "pastoral" aspect of the problem. In their opinion, by opposing the scientific community,

---

24. A heresy may be either formal or material. Material heresy occurs when someone makes a statement contrary to the Faith but is not aware that it is contrary to the Faith. The claim in this case is made by mistake or in ignorance. In contrast, formal heresy occurs when a person makes a claim knowing that it is contrary to the Faith, and does so willingly. In this case, the Inquisition judged that the two astronomical statements were formally heretical, which meant that they could not be taught without the proponent falling into formal heresy.

25. This was the attitude of, for instance, the first "Catholic evolutionists," such Dalmace Leroy, John A. Zahm, and Raffaello Caverni.

the Church alienates many scientifically minded believers and erects insurmountable obstacles that prevent them from accepting the Faith with its message of love and salvation. Thus, in the name of dialogue, peace, and openness to doubting scientists, the Church should be quick to welcome "biological explanations" of origins and should maintain a clear distinction between the authority of the Bible and the authority of science. Yet, by "clear distinction" they mean the actual isolation of theology from any realistic account of origins. The reasoning of these authors can be represented by the following syllogism: The goal of the Bible is not to speak about physical matters. The theory of evolution deals with physical matters. Therefore, the Bible has nothing to do with evolution and for this reason cannot contradict it.

Unfortunately, the authors who think within these lines give into dictatorship of the so-called "scientific community." Their understanding of science-faith relations excludes any biblical authority in answering questions of origins, perhaps with the exception of a few very general statements, such as, "ultimately everything comes from God," "God loves all creation," or "God is present in everything." In what follows, we will show how this popular reasoning misses a few important distinctions.

## Galileo and the Bible

Shortly before his first condemnation, Galileo wrote a letter to Duchess Christina of Tuscany (1615) in which he tried to explain why the Copernican system should never be condemned for religious reasons. There is little doubt that it was this letter that drew the Inquisition's attention to the issue of heliocentrism in Galileo's writings. Pope John Paul II recognized the letter as "a short treatise on biblical hermeneutics." According to the pope, Galileo paradoxically proved better in applying criteria of scriptural interpretation than most of his theological opponents.[26] Let us therefore see what kind of hermeneutical criteria Galileo proposed and how they establish a boundary between the biblical message and the scientific claims of astronomers. Then we will consider whether Galileo's hermeneutics reconciles the biblical message with biological macroevolution or perhaps merely makes room for the scientific understanding of astronomy.

---

26. See the text of the 1992 Address to the Pontifical Academy of Sciences, on the subject "Faith Can Never Conflict with Reason."

The first (of two) principles of biblical interpretation laid down by Galileo reads:

> [T]he authority of the Holy Scripture aims chiefly at persuading men about those articles and propositions which, surpassing all human reasoning, could not be discovered by scientific research or by any other means than through the mouth of the Holy Spirit himself.[27]

Thus, according to the Italian scholar, the revealed knowledge of the Bible is complementary to the knowledge acquired in the natural sciences. Something that cannot be learned by science should be known from revelation, but if science can teach us about a particular matter, then this matter should be the object of natural science rather than revelation. Galileo, however, strongly mitigates the authority of science when he adds:

> Moreover, even in regard to those propositions that are not articles of faith, the authority of the same Holy Writ should have priority over the authority of any human works composed not with the demonstrative method but with either pure narration or even probable reasons.[28]

We see that for Galileo, even those passages of Scripture that concern nature (not just invisible matters of faith) should be first recognized as true based on the authority of the Bible. Later in the letter, Galileo explains this with greater precision:

> [I]n the learned books of worldly authors are contained some propositions about nature which are truly demonstrated and others which are simply taught; in regard to the former, the task of wise theologians is to show that they are not contrary to Holy Scripture; as for the latter (which are taught but not demonstrated with necessity), if they contain anything contrary to the Holy Writ, then they must be considered indubitably false and must be demonstrated such by every possible means.
> So physical conclusions which have been truly demonstrated should not be given a lower place than scriptural passages, but rather one should clarify how such passages do not contradict those conclusions; therefore, before condemning a physical proposition, one must show that it is not conclusively

---

27. Galileo, "Letter to the Grand Duchess Christina," in *The Essential Galileo*, 117.
28. Galileo, "Letter to the Grand Duchess Christina," in *The Essential Galileo*, 117.

demonstrated. Furthermore, it is much more reasonable and natural that this be done not by those who hold it to be true, but by those who regard it as false.[29]

Galileo's position implies that there are two types of statements in the Holy Scriptures. One type is theological and concerns religious and moral matters. These statements are authoritative, independent from science, and constitute the essence of the Bible. Occasionally, however, there are statements in the Bible that concern the natural order, such as the movements of the planets, weather conditions, the biology of animals or plants, etc. Our reading of this latter type of biblical statements should be enlightened by scientific discoveries. This can be done in two ways: (1) If science is not quite sure about some phenomenon, i.e., relies only on—as Galileo puts it—"bare assertions or probable arguments" regarding a given natural event, then we should still understand the biblical passage as it is given, according to the natural meaning of the words and not according to the proposed scientific assertion. (2) However, there are some instances when science can "rigorously demonstrate" how an event happens. *Only then* the scientific conclusion should be applied to the biblical text in order to understand the Holy Scriptures properly.

Now let us apply Galileo's hermeneutical principles to two biblical statements that address the physical universe in order to see how scientific knowledge can enlighten our understanding of the Bible.

The first statement concerns the immobility of the earth and its central position in the solar system. There are a few scriptural passages that were used to defend geocentrism in Galileo's times. One of them is found in Psalm 104:5: "The Lord set the earth on its foundations; it can never be moved" (cf. Ps 93:1, 96:10, 1 Chr 16:30). In another place we read about the movement of the sun: "And the sun rises and sets and returns to its place" (Eccl 1:5). According to Galileo, if a given passage speaks about physical matter it should be understood literally, unless science proves something else beyond reasonable doubt. Although in Galileo's time the movement of the earth and the stability of the sun were still merely hypotheses discussed among scientists, Galileo believed that he could rigorously demonstrate the superiority of the Copernican model over the Ptolemaic one. Even though the earth does indeed move, Galileo's demonstration was not satisfactory. For example, he wrongly thought that tides are evidence of the earth's movement. Thus, we see that Galileo

---

29. Galileo, "Letter to the Grand Duchess Christina," in *The Essential Galileo*, 126.

did not exactly follow his own principle. If he had followed it, he would have waited until a better demonstration of his claim had been presented, and only then would he have come up with a new interpretation of the scriptural passages. Still, the haste that led Galileo into trouble does not diminish the validity of his hermeneutical principle. On the contrary, had he followed it, he would not have got himself into trouble.

Within a couple of centuries after Galileo, it turned out that the movement of the earth can indeed be demonstrated beyond reasonable doubt. Consequently, all scriptural texts suggesting otherwise should be understood not as scientifically describing physical reality but rather as describing physical events in a way that was understandable for the people of the time. Even today, in our overwhelmingly scientific culture, people commonly speak about the movement of the sun rather than the earth. It would sound unnatural or even comical if someone, upon seeing a beautiful sunrise, commented, "What a beautiful motion of the earth that is unveiling the sun before our eyes!" Most people would simply exclaim, "What a beautiful sunrise!" This natural way of speaking was employed also by the biblical authors. When biblical texts (such as the one quoted above) speak about the earth being firmly established, again, they appeal to our common ordinary experience. The earth is typically referred to as a stable surface for human activity, as in the sayings "to have both feet on the ground" or "to be a down-to-earth person." After all, the earth remains our best point of reference for human activity. We see that Galileo's hermeneutical principles enable us to avoid the error of using the Bible to deny facts established by science.

The second biblical statement that we will consider comes from the book of Genesis (2:7): "Then the Lord God formed the man out of the dust of the ground and blew into his nostrils the breath of life, and the man became a living being" (see Gen 2:19, 2:22). According to Galileo, the natural and physical understanding of these and similar words should be abandoned only if something contrary to it is "rigorously demonstrated" by science. Most of the proponents of biological macro-evolution (whether they are theologians or scientists, believers or non-believers) consider the evolutionary origin of man to be virtually certain in the context of what we know about biology. Yet, subjective certitude or even a strong ideological commitment to some idea cannot make up for a rigorous scientific demonstration. The problem is that the origin of man, as well as all other theories of origins, by their very nature cannot be demonstrated in the same way as, for example, the current movements of the planets. Owing to the fact that events related to origins occurred over extremely long periods of time and that the time available for scientific observation is extremely short (relatively speaking), these *historical* theories cannot be proved in the same way as *physical* theories concerning present events. (See our discussion of Diagram 7A and 7B and how moving into the past increases the philosophical and religious component in any theory, at the expense of hard science.) And even if the genetic and paleontological evidence were as sound as the proponents of evolution usually claim, it could not alter the natural and historical sense of the passage from Genesis quoted above. Why? Because even if there were indeed an entire lineage of creatures ascending from apes to humanoids (which is not what we actually find in the scientific data), still the first human being might have been produced exactly as described in Genesis outside and regardless of that biological chain. In other words, the very fact that we cannot directly observe how the first human being emerged makes it impossible for science to disprove his creation from the dust of the earth. For this reason, the evolutionary origin of man can never be "rigorously demonstrated" by science. And this is what Galileo deemed necessary *before* one should abandon the literal sense of a biblical passage. Therefore, according to Galileo's first hermeneutical principle, the Bible can be reconciled with heliocentrism, but not with the evolutionary origin of man.

The second hermeneutical principle Galileo adopted from his contemporary, a clergyman named Baronius, who said, "The intention of the Holy Ghost is to teach us how one goes to heaven, not how heaven goes" (Lat. *Spiritui Sancto mentem fuisse nos docere quomodo ad coelum eatur, non quomodo coelum gradiatur*). A common way of misunderstanding this principle is to think that the Bible does not provide any supernatural knowledge about physical reality and instead is concerned solely with moral teachings and theological truths regarding God and the invisible realm. First-grade naturalists (according to our discussion of naturalism in I,2,b) understand Baronius-Galileo's saying in this way. But this is a reductive approach that implies that the Holy Scriptures do not teach us any worldview or convey any message about nature. On this approach, all metaphysical content in the Bible boils down to ethics and spirituality, or, in other words, metaphysics is reduced to ethics.

This is not what Galileo postulated. To understand his saying properly, we need to carefully clarify what the phrase "how heaven goes" actually means. It does not refer to all supernatural knowledge about the natural order ($t_2$ in Diagram 1). Rather, it refers to the same "rigorously demonstrated" facts of nature that are the object of Galileo's first hermeneutical principle. Galileo means that the Bible will not tell us how the natural world is built or how it works; it will not explain tides, the movements of planets, the growth and habits of living beings, etc. Galileo's saying, however, does not refer to the origins of the universe. The reason is that biblical passages speaking about, for instance, the origin of species do not refer to "how heaven goes" but rather "where heaven comes from."

Evolutionists believe that the origin of species is explained by science. But they rely on evidence for microevolution, postulating that the same evidence testifies to macroevolutionary changes.[30] For instance, population genetics tells us how a population of desert foxes would adapt to polar conditions or how a bacterial strain acquires resistance to antibiotics. These are well-demonstrated and well-understood phenomena, to the point that they can be called "biological facts." Since they explain the ordinary actions of nature, they fall into the category of "how heaven goes." Evolutionists extrapolate these data and say that the same mechanism that explains microevolutionary adaptations of foxes and bacteria can also explain the very origin of foxes and bacteria. In this way, they switch from the question of *how* (how the universe works, how biology and physics work) to *from where* (where biology comes from). Galileo's principle applies only to the first type of question, not to the second type. Therefore, if there is any biblical passage that refers to microevolutionary changes, it should be understood according to what well-established science says. But whenever a biblical text speaks about origins, the passage should be interpreted according to its most natural and literal sense.[31]

If we now apply this distinction to two biblical claims—the stability of the earth and the origin of man—we see that Galileo's principle justifies a non-literal reinterpretation of the former, but it does not allow any metaphorical reinterpretation of the latter. Genesis 2:7 is not about "how heaven goes" but "where heaven comes from" ("heaven" in this case meaning the whole natural order, with all distinct natures, such as the

---

30. For the definitions of "evolution" and "species" adopted here, see II,1.

31. It is a general principle of Catholic exegesis that one should not depart from the literal and obvious sense, except only where reason makes it untenable or necessity requires. This principle first laid down by Augustine was made Catholic teaching by Pope Leo XIII, in *Providentissimus Deus*, no. 15. But in the case of creation (such as the creation of species) it is neither unreasonable nor untenable that they were created separately and directly by God.

nature of man, being a part of it). In any event, the origin of the first man and the first woman, as well as the origin of species, do not belong to the ordinary actions of nature. Rather, they are unique events that happened only once in a distant past. This complies with our Diagrams 7A and 7B, where "now" represents the time of ordinary works of nature guided by unchangeable and uniform laws, whereas "past" refers to the so-called "history of creation," in which ordinary divine providence and laws of nature are not sufficient to explain crucial events. To fully understand what happened in the past, we need supernatural revelation that discloses supernatural divine actions. And this is precisely why Galileo says that "the Bible was designed to persuade men of those articles and propositions which, surpassing all human reasoning, could not be made credible by science, or by any other means than through the very mouth of the Holy Spirit." The origin of species, and specifically the origin of man, is precisely an article that cannot be made credible by science.

Surprisingly, evolutionists who use the Galileo affair to promote biological macroevolution as compatible with the Bible interpret the Bible in a way Galileo himself never would. As Galileo put it, "As to the propositions which are stated but not rigorously demonstrated, anything contrary to the Bible involved by them must be held undoubtedly false and should be proved so by every possible means." Apparently, the idea that man evolved from a lower animal is the proposition "stated" by scientists but "not rigorously demonstrated." It is also clearly in opposition to the Bible, which teaches that man was created out of the dust of the earth. Therefore, according to Galileo himself, the evolutionary origin of man should be held "undoubtedly false" and disproved "by every possible means." Galileo's hermeneutical principles protect the Bible against a reinterpretation that would make it compatible with naturalistic theories, such as biological macroevolution—a theory that goes beyond the limits of science (as has been shown in I,4,c). On the other hand, Galileo's hermeneutics make room for truly scientific discoveries and foster a rational, i.e., non-fideistic, approach to the Bible.

## 6. The Solution to an Apparent Conflict between Science and the Bible

### a. "How?" vs. "From Where?"

The problem that Galileo confronted in his letter to Duchess Christina concerned finding one universal and effective tool for distinguishing what in the Bible can be reinterpreted according to scientific discovery and what should be maintained as religious truth, even if scientists may occasionally challenge it. Finding such a "golden principle" would be truly helpful for the Church. It would facilitate biblical scholarship, help avoid condemnations such as that of Galileo, and secure the biblical message against unfounded claims of scientists with a materialistic mind-set. But does such a golden hermeneutical principle exist at all? In our opinion it doesn't, because this would mean that there is a human idea that could measure God's wisdom contained in Holy Writ. Ultimately, the only way to properly interpret the Bible is by reference to the Holy Spirit, who reveals to the Church which passages speak literally about supernatural events and which are merely making use of human language relevant to a given historical context. This is why no ultimate hermeneutical principle exists. Instead, there is the authority of the Church and supernatural wisdom allowing individuals to interpret the Bible properly.

Having made this reservation, we are ready to formulate an interpretative criterion that—as imperfect as it is—takes us closer to a proper understanding of the Bible, especially with regard to the question of origins.

Before doing so, however, we need to point out some difficulties that could be raised against Galileo's principle. One of them is that his principle is based entirely on the distinction between "bare assertions or probable arguments" on one hand and "rigorous demonstration" on the other. Scientists would say that in science there is constant progress that turns bare assertions into rigorous demonstrations over time. And Galileo would surely agree with this; his principle does not imply anything contrary to it. All he says is that only *after* an assertion has been rigorously demonstrated (to the point that it becomes a "fact") can it be used to reinterpret a passage of the Bible. This order, however, is difficult to maintain. We constantly see that scientists tend to present ideas that are merely asserted as proven. Also, there may be no agreement on whether a given assertion is thoroughly demonstrated or not. When it comes to, for example, biological macroevolution, many scholars maintain it is thoroughly demonstrated while others say it is not. This difference can affect theology and biblical exegesis. Precisely because so many theologians assume that the evolutionary origin of man is scientifically proven, they are inclined to see Genesis 2:7 (and similar passages) as merely metaphors, stories akin to ancient mythologies that influenced Hebrew culture—rather than as historical truth. Thus, Galileo's criterion remains valid, but it does not resolve which statements are mere assertions and which are rigorous demonstrations.

The criterion we are going to present here is inspired by the already quoted saying popularized by Galileo: "The Bible teaches us how one goes to heaven, not how heaven goes." In fact, the Bible teaches us how to go to heaven and where heaven comes from, as well as where heaven is ultimately headed, while it does not say how heaven goes. In other words, the Bible gives us an account of the ultimate things—the first origins of things in a distant past and the ultimate destiny of things in the future. The same applies to humanity. The Bible does not tell us, for instance, how human biology works, but it tells us about the origin of humanity and its destination. Hence, the criterion is based on the distinction between two questions: *from where?* and *how?* If a given scriptural passage poses a difficulty in the context of science, then we need to determine to which question it is more properly directed. If it answers the question *from where?*, then its meaning should not be reinterpreted according to the scientific proposition, but if it answers the question *how?*, then the authority of science must be taken into account, as long as the scientific proposition can be rigorously demonstrated.

Let's refer to some examples to make it clear. Jesus says that the mustard seed is the smallest of all seeds (Matt 13:32). Today's biology can rigorously demonstrate and prove beyond any doubt that there are plant seeds smaller than a mustard seed, for example, orchid seeds. This is a fact. Does it mean that Jesus was wrong or that the Bible teaches something against science? No. It means that in Jesus's time, the mustard seed was the smallest known in His region and that this passage simply has a purpose other than teaching us about the sizes of plant seeds. Jesus employs a metaphor of a growing seed to explain how the heavenly kingdom grows and matures in human history. Jesus's statement is properly attributed to the question *how?* rather than *from where?* His words address the size of the mustard seed, how it grows, how it becomes a shelter for birds. And because it answers *how?* rather than *from where?* it should be interpreted in light of natural knowledge and science, such as biology.

For this reason, no one can claim that the mustard seed is the smallest of all seeds simply because the Bible says so.

This principle applies to all instances in which the Bible speaks about natural events such as planetary movements, weather conditions, animal life, and so forth. In all these instances it is not the goal of Scripture to explain natural phenomena. Yet, there are other passages where the things work differently. For example, when Luke presents the genealogy of Jesus, he ends it with the statement, "Enos, [who was] the son of Seth, the son of Adam, [who was] the son of God" (Lk 3:38). In this case, the Evangelist answers the question *from where?* rather than *how?* After all, the genealogy of Jesus is included in the Gospel precisely to explain where Jesus's humanity came from. We see that Adam is called "the son of God." Reason alone tells us that God does not produce children in the same way as people do. Therefore, it is obvious to any reasonable reader that this passage does not say that Adam was born from God in the same way that Seth was born from Adam. Nevertheless, we cannot reinterpret this passage based on alleged "scientific evidence" because it addresses a question of origins. Thus, this verse should be understood according to its theological meaning, which is that Adam was made directly by God, without any secondary (intermediate) causes involved. There is not any medium mentioned between Adam and God—Adam descended immediately from God, that is, he was made directly by God. This is how the human family came into existence, and then it began to naturally propagate. The same principle of interpretation should be applied to many other passages of the Bible where science seems to conflict with theology.

A number of objections may be raised against the hermeneutical principle proposed here. We will address at least those that result from a confusion that could easily occur in this context. One of the "confusion-type" objections can be formulated like this: If biblical statements addressing *how?* are susceptible to reinterpretation in the light of science, then most miracles should be invalidated. The reason is that miracles tell us *how* a thing happened and not *where* it comes from. A good example is the multiplication of the loaves. Jesus kept sharing five loaves of bread until he satisfied more than five thousand people. The whole event falls into the category of *how* (how Jesus fed five thousand, how bread was multiplied). Naturalists and atheists believe that if this event actually happened, it must have a natural explanation, even though science cannot yet produce any such explanation. We should not conclude, therefore, that this event involves any supernatural action. With this reasoning, our hermeneutical principle fails because it allows us to explain away virtually all miracles recounted in the Bible.

To answer this objection, it is enough to check what the Bible itself says. In this case, the Bible states that the multiplication of the loaves was a miraculous event (cf. Jn 6:14). Most miracles performed by Jesus and the Apostles are explicitly called miracles in Holy Scripture, so there is no need for additional interpretation here. This approach complies with a greater hermeneutical principle saying that one scriptural passage should be interpreted in the light of others. In contrast to the *how?* questions, issues that fall into the *from where?* category must refer to a distant past. It is safe to say that they need to involve the origins of the universe and humanity, and not the later operation of the physical universe or the history of salvation (i.e., the area near now in Diagram 7B). Hence, our principle does not apply to the interpretation of miracles that happened in Jesus's times or human history.

Interestingly enough, scientists of the highest stature, before naturalism came to dominate Western culture, understood the distinction between *how?* and *from where?* quite well. Neither Copernicus, nor Galileo, nor Kepler, nor Newton addressed questions of origins in their scientific writings. Newton leaves us a clear evidence of this distinction in one of his works:

> Though [celestial bodies] may, indeed, continue in their orbits by the mere laws of gravity, yet they could by no means have at first derived the regular position of the orbits themselves from those laws . . . [Thus] this most beautiful system of the sun, planets, and comets, could only proceed from the counsel and dominion of an intelligent and powerful Being.[32]

Newton believes that there is no need for a supernatural activity of God to maintain planetary movements. This happens thanks to gravity. But to set up the planetary system in the first place belongs to the divine mind.

Our distinction becomes even more clear when we compare the titles of the two most influential books in modern science—Copernicus's *On the Revolutions of the Celestial Spheres* and Darwin's *On the Origin of Species*. Copernicus's book treats the movements of planets, and thus it falls into the category of *how* the universe works. His conclusions, therefore, if they are well established and rigorously demonstrated, should

---

32. Isaac Newton, in Florian Cajori, ed., *Sir Isaac Newton's Mathematical Principles of Natural Philosophy*, 543–44.

shed light on our understanding of biblical passages, specifically those presenting the earth as an immobile planet and the sun as a body in motion. But with Darwin it is different. The British naturalist employed the term "origin" in the title of his book. The English word "origin" stands for Greek word "genesis," which is the word used in the Septuagint as the title of the first book of the Bible. Genesis was intended to explain the origins of things—of the universe, species, and the human race. And this also was Darwin's goal—to explain the origin of species and man. Darwin thus violated the limits of natural science, and this is why his explanations conflict with those found in Genesis. There is not much exaggeration in saying that Darwin came up with an alternative Genesis—one that is necessary to make the materialistic worldview coherent, yet one that cannot be used to reinterpret the Bible.

※

Until now, we have explained why scientific theories fail to overturn the literal and historical interpretation of Genesis inasmuch as it addresses the problem of origins. Our defense of the supernatural character of Genesis is based on the distinction between two types of questions addressed by biblical texts—the question of *how* things work and *where* they came from in the first place. Nevertheless, there is still one more serious objection that could be raised against our distinction.

In older theology, the origin of geological formations (specifically mountains) was attributed to direct, supernatural divine causation. Later, however, it turned out that natural processes satisfactorily explain these phenomena. Does this not completely ruin the distinction between *how?* and *from where?* If the origin of mountains and valleys can be attributed to secondary natural causation, such as the shifting of tectonic plates, why shouldn't we attribute the origin of living beings to secondary natural causes, such as random variation and natural selection? In fact, a great number of contemporary scholars adopt a way of thinking based on the extrapolation from geology to biology. (Indeed, this was also a thought-pattern employed by Darwin.) This seems to ruin our distinction and hermeneutical principle because apparently both the origin of geological formations and the origin of species fall into the *where from?* category rather than the *how?* category. Since we found a natural explanation (natural secondary causes) for different geological forms, at least some biblical statements attributing directly to God the origin of some elements of the universe cannot be understood literally. If this is the case with mountains, then why should anything else be made directly by God? To answer this objection, we need to elaborate on this issue by providing greater detail and some historical background.

## b. The Bible and the Origin of Geological Formations

In 1566, Pope Pius V published the first Catechism of the Catholic Church. Catechisms by themselves are not a source of Church teachings. Rather, they are a summary of the Catholic doctrine contained in conciliar decrees, papal pronouncements, and other expressions of the teaching office of the Church (*Magisterium Ecclesiae*). Therefore, if a catechism does not present Church teaching accurately, we say it contains an error, even though Church teaching itself is infallible. Since catechisms include both solemn and ordinary doctrines, they are probably the best sources for understanding how the Catholic Church approached some secondary theological problems in a given era. In the Roman Catechism of 1566, we read the following about creation:

> The supreme Architect . . . created all things in the beginning. He spoke and they were made: He commanded and they were created . . . The earth also God commanded [*verbo suo iussit*] to stand in the middle of the world [*in media mundi parte*], rooted in its own foundation, and made the mountains ascend [*effecit ut ascenderent montes*], and the plains descend into the place which He had founded for them. That the waters should not inundate the earth, He set a bound which they shall not pass over; neither shall they return to cover the earth. He next not only clothed and adorned it with trees and every variety of plant and flower, but filled it, as He had already filled the air and water, with innumerable kinds of living creatures. Lastly, He formed man from the dust of the earth, so created and constituted in body. By referring to the sacred history of Genesis the pastor will easily make himself familiar with these things for the instruction of the faithful.[33]

We see clearly that the views presented here are derived not from science but from the Bible. The last statement makes it explicit that it is Genesis, not studies of nature, which makes us familiar with the origins of the universe. Therefore, what we read in the catechism is a consequence

33. *Catechismus ex Decreto Concilii Tridentini*, 22–23.

not of the "bad science" of the time but of a more fundamental conviction—namely, that the origins of the universe cannot be explained by science. Instead, they must be derived directly from theological sources, such as the Genesis account.

Precisely this principle seems to be challenged by later scientific discoveries. The catechism implies that several features of the universe were made supernaturally (and perhaps directly) by God: God utters the word, and the earth rests in the middle of the universe; God utters the word, and mountains are elevated; and the same with different living beings. Even though some scientists think otherwise, we do not have any well-established theory explaining the origin of the solar system and the earth. For now, we are leaving aside the question of the origin of the earth and focus on the geological processes that are relatively well known to us thanks to modern geology.

When it comes to the mountains, we know that they emerge as a consequence of movements of tectonic plates and some other mechanisms (erosion, volcanic activity, etc.), all of which geologists have explored thoroughly. But we also learn from the Bible that at some distant point in the past all the earth was covered with water. Science cannot disprove this biblical message, but perhaps it could confirm it. If all the earth was covered with water, most probably earth's crust was in perfect balance; tectonic plates most probably did not even exist. Hence, the earth must have transitioned from the stage of some kind of geological balance to the stage of tectonic plates dance that takes place ever since. Science can speculate on what could cause such a transition (meteor impacts, lunar gravitation, etc.), but it cannot exclude the possibility that at this point God worked supernaturally, either via supernatural secondary causes (such as angels), or directly, by causing changes that started the formation of our current mountains and continents. In any case, science cannot exclude God's supernatural activity at the very beginning of geological processes. Hence, the statement of the old catechism could be essentially correct: God indeed caused the mountains to ascend and the plains to descend, even if these were not the same mountains that we, or the authors of the catechism, had in mind. In other words, the mountains we know have been made naturally, but the first geological formations might have their origin in some sort of supernatural causation.

In the case of mountain formation, all Christianity—from antiquity until the late eighteenth century—believed that God formed mountains, valleys, and shorelines more or less in their current form. This was a somewhat naïve picture if we judge it based on our current knowledge. Modern geology has changed this view by showing there were many natural processes that influenced the current appearance of the earth. But the same science has not disproved and cannot disprove any and all forms of supernatural causation at the very beginning of these processes. Hence, the distinction between *how?* and *from where?* remains valid: geology tells us *how* the mountains are formed as we know them today, but the same geology cannot tell us *where* the geological formations came from in the first place. At the very least, science cannot prove that there was no supernatural action whatsoever on the part of God at the beginning of geological processes.

At the same time, we need to notice that mountains, valleys, or river beds are not any essential elements of the earth or the universe. The Bible does not teach us about the origin of things that are not quite distinctive, rather it explains how some essential elements or entirely new natures came about (such as animals and plants). And even this is done in a general way without giving much detail. Geological formations are not some distinctive forms or "separate natures," and therefore their formation might have been performed by natural causes. Moreover, it seems that the editors of the 1566 Catechism did not exactly follow the biblical account. Genesis does not tell us anything about the elevation of the mountains. It only speaks about "gathering the waters," which may refer to the creation of the pre-continent (called the "dry land"). The message about the formation of mountains is contained in other scriptural passages, such as Psalm 104:6b–9 and 95:5.[34] But these other passages are not designed to explain the origins and thus they do not need to be read according to the historical and literal mode.[35] Genesis reveals that there was

---

34. Psalm 104:6b–9: "Above the mountains stood the waters. At your rebuke they took flight; at the sound of your thunder they fled. They rushed up the mountains, down the valleys to the place you had fixed for them. You set a limit they cannot pass; never again will they cover the earth." Ps 95:5: "He made the sea; it belongs to Him, the dry land, too, for it was formed by His hand."

35. For example, a Decree by Pontifical Biblical Commission from June 23, 1905, *On Narratives Historical Only in Appearance in Books of Holy Scripture Historical in Form*, establishes that a historical sense cannot be doubted in the case of the biblical books that are considered historical in their character. (And even then, the Commission makes some reservation.) But the Book of Psalms is not considered historical in its form therefore it is possible to interpret the Psalms in a more figurative or metaphorical way. Also, the 1909 Decree *On the Historical Character of the First Three Chapters of Genesis* does not list the formation of the mountains among those truths that have to be considered historically and literally true.

some kind of supernatural activity on the part of God in the gathering of waters (Gen 1:9). This may or may not refer to the beginning of the geological processes. Since Genesis is not explicit, there is no strong biblical evidence for direct divine activity in the beginning of geology. The role of God in the first formation of the earth's surface is hard to determine. At this point it is fair to say that, most probably, there was some kind of supernatural work of God in the first formation of the earth's surface (between establishing the earth and the production of living beings) but we cannot resolve whether or not it entailed the direct divine causality and we cannot tell what exactly was caused by that divine work.

Based on this we can conclude that the case of mountain formation does not ruin our distinction between *how?* and *from where?* because in this case we do not speak about any distinctive entities (new natures), so their origin cannot be clearly defined. In other words, it is not clear whether the origin of mountains fall into the category of *how?* or *from where?* Moreover, there is no decisive biblical argument for the supernatural formation of geological entities, so this mode of their origin cannot be conclusively established even if we strictly follow the literal and historical sense of Genesis.

Our answer can be put concisely as follows: the case of the origin of mountains does not ruin our criterion but rather escapes it, because it is not clear to which category (*how?* or *from where?*) it belongs and whether the Bible teaches their supernatural origin or not.

### c. The Progress and the Limits of Science

If we step out of the problem of mountain formation and adopt a broader perspective, we realize that there were a few other scientific discoveries that apparently challenged the traditional understanding of the creation story as presented, for example, in the 1566 Catechism. Let's refer to a few other examples.

That the earth is a globe had been known since antiquity. But let's imagine that some uneducated people thought the earth was a flat disc. This was probably the case with the ancient Jews, who were the first recipients of the biblical message. For them, the Bible was compatible with the idea of a flat earth sitting still in the middle of the universe. To this common view the book of Genesis added the age of the universe, that is, roughly six thousand years. But later, astronomy and physics proved beyond any doubt that the shape of the earth is approximately spherical. This is a fact. Hence, science modified the worldview of the ancients and influenced biblical interpretation. Science, however, has never disproved the essential message about the origin of the earth, namely, that God established the earth by some special action.

The second great challenge came with Copernicus, who postulated the mobility of the earth and the stability of the sun as the center of the solar system. This new scientific discovery caused turmoil and significant disagreement in the Church culminating in the controversy about Galileo. However, within roughly two centuries this problem was overcome, and again it turned out that none of the truths of the Faith were challenged by the heliocentric model of the solar system. Only a few scriptural passages needed to be understood differently, that is, as a natural human way of speaking about celestial phenomena, rather than a lecture on astronomy.

Later, the age of the earth and the universe turned out to be much greater than biblical genealogies suggest. Again, many Christians did not agree with this discovery and tried to defend the literal accuracy of the genealogies and the idea that the six days of creation were six natural days. This approach still exists in the form of young-earth creationism.

Right after the discovery of deep time (i.e., the concept whereby natural history extends billions rather than thousands of years into the past), Charles Lyell came up with his geological theories that established natural laws governing the formation of mountains and valleys. Still later, the idea of fixed stars was challenged, and it turned out that our solar system does not sit in the middle of the universe but rather is located midway from the center of one galaxy among billions. On top of that, it turned out that there are many planetary systems similar to ours.

Today, many scholars who are familiar with the history of science (and how it has modified our biblical worldview) believe that this process has no limits and never ends. As a consequence, they think that the origin of the universe and everything in it should be explainable without any reference to supernatural divine causation. This perspective was adopted even by the theologians of the International Theological Commission.[36] Considering all these facts, how could anybody possibly defend the special creation of man or the separate creation of species?

To answer this question, we need to again return to the distinction between *how?* and *from where?* All theories that legitimately modified the

---

36. See the 2006 document of the International Theological Commission, *Communion and Stewardship: Human Persons Created in the Image of God*, nos. 62–63.

Christian understanding of the Bible address the question of *how?* They refer to how the universe works, how it is built, but not where its elements came from in the first place. Thus, the fact that the earth is a globe tells us what the earth looks like and what its shape is. The fact that the earth moves tells us about the solar system, how it is built, what kind of laws govern it, what are the relative positions of the planets, etc. The fact that there is nothing like a fixed sphere of stars, but instead billions of galaxies forming larger spatial structures, tells us how big the universe is, how its matter is distributed, and what kind of celestial bodies and phenomena there are. Again, this refers to "now," not to origins. Even the age of the universe is a question of "now"—it does not tell us *where* the universe came from, but *how* old it is now. And this is why the age of the universe is a legitimate scientific problem that should be settled by science, not by the Bible. In contrast, the theory proposed by Darwin addresses the question of *where* the different natures of living beings came from in the first place. Since Darwin's theory sets a goal which is fundamentally different from those of typical scientific theories, it should not be adopted to reinterpret traditional Christian approach to Genesis. Table 1 contains a summary of different theories and their division into the two groups (*from where?* and *how?*).

| | Theories of | | | | |
|---|---|---|---|---|---|
| | Nature (how?) | | Origins (from where?) | | |
| | Theology (the Bible) | Natural Science | Theology (the Bible) | Natural Science | |
| Solar system | Geocentrism | Heliocentrism | Supernatural action of God (direct or indirect) | The gravitational collapse of a region within a large molecular cloud that caused faster rotation and formation of the planets | Solar System |
| Shape of the earth | The Bible itself suggests globosity; ancient Hebrew culture accepted a flat earth | Approximately spherical | Shaped by God (directly or indirectly, not in the current form) | Shifting of tectonic plates | Geological formations |
| Model of the cosmos | Fixed stars above the earth | Billions of galaxies distributed unevenly throughout an immense space | Supernatural formation | Biological macroevolution (based on the neo-Darwinian mechanism) | The origin of species |
| An age of the universe | About 4 thousand years before Christ | 13.8 billion years | Direct formation out of the dust of the earth and infusion of the soul | Descent from a lower animal through random variation and natural selection | The origin of man |
| | Secondary discipline (if it contradicts the primary discipline the primary should be followed) | The primary discipline for establishing the truth | | Secondary discipline (if it contradicts the primary discipline the primary should be followed) | |

Table 1. Explanations of different natural phenomena with regard to their operation and origin. Questions of *how?* are addressed primarily by science, whereas questions of *from where?* belong primarily to theology. Accordingly, answers to the first type of question should be established by science, and answers to the second type by theology.

The table does not contain the scientific theory of the Big Bang or the theological concept of *creatio ex nihilo*. This is because these theories do not overlap. The Big Bang concerns the early development of the universe (beginning with the one-Planck-old universe), whereas *creatio ex nihilo* concerns the very origin of the universe. Since they address two different problems, there is no conflict between them, but rather perfect compatibility. We must keep in mind that Big Bang cosmology, in principle, does not exclude divine guidance of the early expansion of the universe. In fact, the inflation model (the idea that the total amount of matter and energy increased in the early stage of universe expansion) harmonizes with the theological concept of the supernatural formation of the universe at the early stages of its creation.

The problem of Darwin's theory is analogous to the problem of pagan beliefs about the eternal existence of the universe. Science and philosophy cannot disprove the possibility of an eternal existence of the universe (in whatever form it would have existed). Before the ever-continuing expansion of the universe was discovered, for centuries materialists had claimed that the universe was eternal. In their minds, an eternal universe does not require any additional causal explanation and thus it is compatible with the atheistic worldview. Ancient pagan philosophers such as Aristotle and Plato were not materialists or atheists, but they did believe in the eternal existence of material reality along with God. This idea was probably more compatible with a simple, nonscientific observation of nature. For those who did not know Genesis, it was certainly easier to accept the eternal existence of matter than creation out of nothing. Thus, we can say that scientific/philosophical theories generally favor the idea of an eternal universe. Yet, Christians have always believed that the universe came from nothing at the beginning of a finite past. Does this mean that Christian belief is antiscientific, or that the Bible should be reinterpreted according to dominant scientific claims? No. On the contrary, Christians have always defended the doctrine of temporal creation, because it is revealed in the Bible and science has never disproved it. The same applies to the formation of the universe—neither the ancient philosophers nor contemporary evolutionists acknowledge the direct formation of essential elements of the universe. Nevertheless, for the same reason that Christianity adopts the creation of the universe out of nothing in the beginning, it should also defend its supernatural formation.

Darwin was able to show many examples of the evolution of species. He failed, however, to show a single example of the origin of a new species (for our definition of evolution and species, see II,1). Since Darwin, many new examples of beneficial adaptations have been described in the scientific literature, but never an origin of a new natural species. Hence, Darwin's theory is based on extrapolation of what he observed in nature into what he assumed about nature. In other words, Darwin assumed that the same biological laws that account for adaptation within species explain the origin of species, and ultimately all biology. Yet, the reality is different. The laws of biology can explain how biology works, but not where biology (i.e., the entirety of life forms) comes from; thus, they cannot explain the origin of species. To say that once we have a living cell the emergence of all biodiversity can occur thanks to the laws of biology (such as generation, variation, and natural selection) is not satisfactory, because as long as we do not have biodiversity we do not have biology, and as long as there is no biology, biological laws cannot operate. One living organism does not constitute the entire biology. Hence, even if one or a few organisms where somehow given in the beginning, still the existence of all biology would need to be explained. And this could not be done by referring to the very laws of biology.

To show this more clearly, we need to refer to a few similar examples. The laws of chemistry, for instance, explain how different compounds behave and react under different conditions. Yet, these same laws of chemistry will not explain the origin of the particles and the properties that are necessary for chemistry to work. Chemistry, therefore, does not explain the origin of chemistry. Since the chemistry of compounds is most probably determined by the physics of particles, a different science that can probably explain the origin of chemistry is physics. But physics does not explain the origin of particles or physical laws. Therefore, physics does not explain the origin of physics. Similarly, astronomy does not explain the origin of astronomy, mechanics cannot explain the origin of mechanics, and mathematics cannot explain the origin of logical rules and numbers. Finally, science as such cannot explain the origin of science, because in order to learn about physical reality, physical reality must be already present. The laws of any discipline can operate only when a relevant object is already available to them. In the same way, laws of nature taken together can operate only when nature is already present and formed in a way enabling the operation of these laws. This is why none of the particular scientific disciplines, or science as such, could ever explain the origin of the universe. Claiming otherwise would amount to saying that a given physical reality is an ultimate cause of itself, which is never the case with any creatures.

The mistake of Darwin consisted of his desire to explain the origin of all biodiversity by recourse to laws operating only when biodiversity already exists. In this sense, his postulates went beyond what was attainable within his own field. And because the Bible speaks about the origin of species, we have no other way to learn about their way of becoming than by adopting biblical message as historically true. As Galileo put it, "The Bible was designed to persuade men of those articles and propositions which, surpassing all human reasoning could not be made credible by science, or by any other means than through the very mouth of the Holy Spirit."

## Chapter I—Summary

In the first chapter we learnt that religion and science have their independent fields that come close to each other when addressing questions of origins. However, science cannot provide an ultimate answer to such questions. Modern science has recognized its own theoretical limits. Wherever science ends, the further inquiry belongs to philosophy and when philosophy encounters its limits, then theology comes to aid. Trespassing the limits of science means falling into reductive philosophy of naturalism. Even though science is amazingly effective in answering the questions about nature—how it works—it cannot explain the first origin of these structures or the ultimate destination of nature as such. To know nature in its entirety we need to resort to the knowledge coming from outside of nature, which is the supernaturally revealed truth found in theology. Theology is the proper discipline to address the questions of origins and destination.

# Chapter II

# Evolution and Natural Knowledge

THE GOAL OF THIS chapter is to explain what natural knowledge—coming from both philosophy and science—tells us about biological macroevolution. Our discussion moves from the relation between religion and science to the problem of one theory of biological macroevolution presented in science, and how this theory confronts our natural knowledge derived from logic and observed facts. The crucial point is to properly define the notions used in the debate over evolution. In fact, the outcome of this debate heavily depends on how we understand the terms. For this reason, we will devote Section 1 to clarifying terminology that we have been using and will continue to use consistently throughout this book. Next, we will discuss some problems for biological macroevolution coming from philosophy—first the logical problems (Section 2) and then the metaphysical problems (Section 3). In Section 4 we will focus on scientific (biological) problems of biological macroevolution commonly recognized by the critics of Darwin's theory. The last section will introduce the basic tenets of the alternative theory in science, i.e., intelligent design.

## 1. Preliminary Definitions

### a. Species

Our discussion thus far has involved two notions that may mean different things for different people. These are *evolution* and *species*. Now it is time to explain how we understand these terms in our context so that a confusion over the notions would not create a disagreement over the conclusions.

When we look into the fossil record, we see that in the early stages of earth's existence life was scarce and biodiversity very limited. Only later do we see more and more new forms of plant and animal life. Regardless of whether we consider the fossil record smooth and continuous or rugged and twisted, we see progress in the biosphere from uniformity and simplicity to diversity and complexity. There is little doubt that over time something new came about that had not existed before. Darwin wrote a book on the origin of species, thereby indicating that for him the novelty emerging in natural history can be defined as new "species." We will see later, though, that Darwin does not clearly define what he understands by "species," and this failure to establish clear definitions leads him into logical problems (see II,2). To avoid confusion in any part of this book, we need to establish what we mean by the emergence of new species.

The most common understanding of species is the modern idea of a *biological species*. This is usually defined as a group of individuals that can reproduce and yield fertile offspring in natural conditions.[1] Thus, the common denominator for biological species is reproductive power and fertility. There are, however, examples (Darwin himself found a few of them) when fertility does not constitute a good criterion for distinguishing between different groups of animals or plants. This is either because some organisms seem to interbreed only under certain conditions, or because they can interbreed under most conditions, but they produce infertile offspring. Based on these "in-between" examples, some biologists would claim that new species may arise in the course of natural (or artificially enhanced) reproduction. Therefore, if *species* means *biological species*, then evolution understood as common descent and the emergence of new species has been demonstrated both in laboratory and in nature.[2] This, however, is not the type of novelty whose emergence the

---

1. Typically, this notion is traced back to *Systema Naturae* (1735) by Carl Linnaeus. The Swedish naturalist was the first to establish modern taxonomy based on five levels of division: kingdoms, classes, orders, genera and (systematic) species. Within species he speaks of varieties in botany and subspecies in zoology. The modern concept of biological species (which corresponds to Linnaeus's *systematic species*) is typically attributed to Ernst Mayr. See his *Systematics and the Origin of Species*.

2. Michael Behe discusses a few examples of the emergence of many new biological species in a relatively short time. See *Darwin Devolves*, 143–70.

theory of evolution attempts to explain. In order to explain the origin of all biodiversity, we need a category broader than that of biological species.

There are also biologists and philosophers who skeptically claim that we can never define what a species is. They say that the notion of species is quite arbitrary and changes according to many possible definitions. This kind of approach stems from the nominalistic understanding of biological classification. Scholars who adopt this attitude try to build general notions without allowing anything like the *nature* of a thing. They attempt to describe groups of objects based on their external features alone (physical features) and only by reference to *how* things are, never asking *what* they are. Since the idea of species consists of the empirical and the speculative (abstract) element, in the skeptical approach the speculative element is ignored or rejected as arbitrary or subjective. And because on strictly empirical level every individual differs from another individual, in the nominalist approach there is a tendency to reject the idea of species other than biological species.

The nominalistic approach assumes the notion of species that we can call a *logical species*. Species in this sense is defined as a logical subcategory of the broader category of genus. Individuals may be included in logical species quite arbitrarily because there is nothing inherent in them that determines their belonging to one or another category. Those who claim that "species do not exist" understand species only in the logical way, not in any realistic or metaphysical sense of the word. It's true that if species did not exist in nature, or their definition was entirely impossible, the discussions on the origin of species would be limited to biological species. Then, the theory of the origin of new (biological) species by random variation and natural selection would be provable (at least to some degree). But such limitation of the notion of species would be like setting up the rules in such a way that one cannot lose the game. It seems that a big portion of "species skepticism" is motivated precisely by the desire to make Darwin's theory demonstrable. As we will see later, Darwin's own attitude includes this kind of skepticism.

But it is not possible to consistently maintain species skepticism in real life. Even skeptics themselves activate their skepticism only when it comes to theoretical considerations, such as biological classification or discussions about evolution. They do not apply their criticism to the ordinary experience of biology. When they get home after a heated debate in which they defended the "nonexistence of species," they do not have a problem with distinguishing their cat from their dog or identifying different species of fish in a neighbor's pond. In fact, by their refusal to acknowledge the real existence of species, skeptics do not make a discussion on the origin of species impossible. Instead, they simply refuse to participate in a legitimate discussion that other scholars continue to have, based on other-than-logical definitions of species.

The logical species constitutes the basis for other definitions of species. Any realistic definition of species is the logical one, supplemented by some elements derived from reality. Thus, in a sense, logical species is the broadest understanding of species, but, at the same time, it is the least meaningful in the debate owing to its detachment from reality. The most abstract definition of species that includes both the logical and the real meaning is that of *metaphysical species*. A metaphysical species includes all individuals that share the same nature or substantial form, regardless of whether they are living beings or any other substances.

The concept of species relevant in the debate over evolution is broader than the biological species but less abstract than the metaphysical. Scholars of the past adopted the traditional idea of the so-called *natural species*. The rejection of the concept of natural species in modern biology causes much confusion. One example of such confusion is the controversy about *species fixism*. According to fixism, the number of species and their appearance have not changed since the beginning of the universe. This kind of species fixism was developed in modern times by Carl Linnaeus,[3] and soon afterward it was challenged by scientists such as Georges-Louis Leclerc (Comte de Buffon). Today, it is a fact demonstrated by evidence that the number of species has changed over time. Still, we can imagine a moderate version of species fixism which assumes that the *number* of species may change over time, but their *appearance* (i.e., the basic form) does not change. The problem, however, is that some scholars, when speaking about the stability of species, have natural species in mind, whereas other scholars think of biological species or even varieties. Hence, both parties may be right, because species

---

3. Linnaeus believed that *Species tot numeramus quot diversae formae in principio sunt creatae* ("There are as many species as there were different forms created in the beginning.") This often-quoted phrase by Linnaeus is followed by the recognition of "secondary causes" after creation was completed: *Variationes tot sunt, quot differentes plantae, ex ejusdem speciei semine sunt productae* ("There are as many variations as different plants are produced from the seed of the same species.") See *Fundamenta botanica*, 18; cf. *Philosophia botanica*, 99. We see that Linnaeus had no problem with recognizing the fact that sometimes new species emerge in the course of hybridization. See "De Peloria," an extract from his larger work *Amoenitates Academicae*.

taken as broader categories (e.g., natural species) may not change even if the narrower categories (e.g., biological species) do change. The fossil record shows changes in the number of species, but it does not disclose changes in their basic forms. Some forms simply appear, and others disappear. Therefore, even today, according to our best knowledge about the history of life, *moderate* species fixism is defensible as long as species is understood broadly enough, for example, as natural species.

Denying the existence of natural species stems from the same spirit of nominalism that leads some skeptics to reject any idea of species. However, the notion of natural species is not as elusive as the skeptics may conceive. It may be defined as a broad taxonomical category, corresponding roughly to our modern biological classification of "family." Thus, it is a broader category than modern biological species or the Linnaean systematic species. For instance, the family *Felidae* includes all kinds of cats: domestic cats, tigers, panthers, lions, jaguars, leopards, and other genera that include many (biological) species. The family *Canidae* includes all kinds of dogs, such as domestic dogs, wolves, foxes, jackals, and coyotes. Similarly, the family *Elephantidae* includes all species of elephants and mammoths. The meaningful debate about evolution is concerned not with the small differences between biological species but with the origin of new families (the differences between, for example, an elephant and a dog or a giraffe).

Natural species can also be defined on levels of knowledge other than the scientific (i.e., biological) one. In philosophy, natural species are living beings that share the same nature or substantial form. In theology, natural species can be identified with the biblical kinds mentioned in the book of Genesis. In Diagram 1 (see I,1,a), the understanding of natural species in different disciplines of knowledge is represented by the arrows $t_2$, $t_3$, and $t_4$. The notion of natural species, whether considered in science, philosophy, or theology, always refers to the same physical reality. Natural species provides a common point of reference for the different levels of knowledge and thus makes the notion of species understandable when switching from one discipline to another. This is why truly interdisciplinary debates on the origin of species are possible. Consequently, whenever in this book we refer to species without qualification, we mean natural species. Diagram 8 presents a summary of our typology for the concept of species.

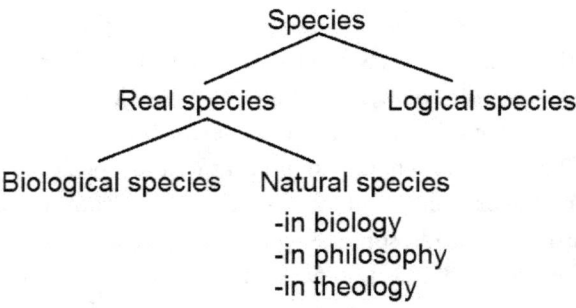

Diagram 8. Division of species.

### b. Evolution

Darwin invented his theory of evolution to explain the origins of biological novelty over time. This, however, was not the case with the ancient and medieval way of understanding evolution. Before Darwin, the term did not mean the emergence of anything new. According to Augustine and other Christian writers, it was just a revealing or manifestation of something already existing in a hidden form.[4] We can say that over the last two centuries the meaning of the word "evolution" changed from *manifestation* (*disclosure* or *unraveling*) of something hidden to *creation* of something new. This is why Augustine and Darwin, when speaking about evolution, have something different in mind—two substantially different ideas of evolution. Darwin himself was aware of the traditional meaning of the word, and for this reason he did not use it until the later editions of *The Origin of Species*. Only when the term gained the new meaning, under the philosophical influence of Herbert Spencer, did Darwin decide to introduce it into his own writings. This was a substantial change because in the traditional sense of the word, evolution means something like "change over time" or simply "development," whereas after Darwin, evolution started to describe the emergence of functional biological novelty.

In traditional Christian theology, the emergence of new species understood as completely new forms of life (i.e., new natural species)

4. See Gilson, *From Aristotle to Darwin and Back Again*, 50–52.

could be attributed only to the direct power of God. After Darwin, it was attributed to natural processes generally referred to as biological evolution. This is the change in the meaning of the word that we should keep in mind whenever we compare Augustine's ideas (and other Christian writers') with the modern concept of evolution (for more see III,3,b).

We have previously used the term *biological evolution*. The word *biological* in this context narrows down the scope of the possible applications of the word "evolution" in different disciplines. Astronomy, for example, speaks about the evolution of stars, and the social sciences speak about the evolution of culture (e.g., language, religion, morality, laws). We are not concerned here with any use other than the one pertaining to biology. Therefore, even if evolution may be a fact in other disciplines, in biology it can remain a theory or merely a hypothesis. Moreover, whatever we say about evolution in biology may be either completely true or entirely false outside of biology.

This distinction is important, because it is not rare to hear an argument of this kind: "Everything in the universe evolves: stars, planetary systems, the environment, human civilization. Life evolves in the same way; therefore, the emergence of new species may be rightly attributed to the general process of evolution that we see everywhere in the universe."

The problem with this kind of reasoning is that the same word—"evolution"—is used with different meanings. If the emergence of a new star, its burning over billions of years, and its final explosion is called evolution, then the counterpart in biology would be nothing more than a succession of generations: a cat bears a cat that from the form of an embryo turns into a fetus, then is born, matures, gets old, and dies. No emergence of any novelty in the Darwinian/Spencerian sense is observed in this process. Similarly, there is no real novelty generated in so-called stellar evolution. For this reason, these natural processes of "change over time" should not be confused with evolution in biology, especially the sort of evolution that is supposed to explain the emergence of new functional organs, new species, or new genes. Evidence from one discipline does not constitute evidence for another—"stellar evolution" is not evidence for the emergence of new species in biology.

For the purposes of this book, we limit the notion of evolution to biology. It does not follow, though, that notions of evolution in different disciplines are completely unrelated; comparing ideas of evolution from different disciplines simply goes beyond our subject matter.

One more distinction is crucial for a proper understanding of the concept of biological evolution. It is the distinction between *micro*evolution and *macro*evolution. We will use "microevolution" to refer to the production of new species or genera. In contrast, "macroevolution" will refer to the beginning of new taxonomic families and higher levels of classification.[5] As we said, in contemporary biology the evidence for microevolution is often extrapolated to serve as evidence for macroevolution. This methodology, however, is questioned by scholars who show that the differences between families are not merely quantitative but qualitative. This means that interfamily transitions would require not just strengthening or developing some features of an organism (usually while weakening others) but also generation of entirely new genes and proteins (on the molecular level) or organs, organ systems and body plans (on the organismal level). According to the critics of the argument based on extrapolation, biological mechanisms are known to generate microevolutionary changes, such as a change in fur color or the size of a beak, but they are not capable of generating macroevolutionary changes, such as different respiratory or circulatory systems, new organs, or even new senses (e.g., the echolocation of a bat).

One way to extrapolate the evidence of microevolution in order to argue for macroevolution is to challenge the very distinction between the two. Some scholars maintain that we cannot speak of micro- and macroevolution because we do not really know where the limits of species are. They usually show some examples of an observable microevolutionary change that seems to alter an organism (or a population) beyond the level of biological species. In response, we acknowledge that microevolution in some cases may go beyond biological species, reaching the level of genus. However, none of the examples provided by Darwin or his followers demonstrate the emergence of new families.

Sometimes Darwin's proponents challenge the distinction between micro- and macroevolution by showing examples of organisms that are harder to classify; for example, they may belong to one or another family.[6] But this kind of challenge does not ruin the division between

---

5. The terms are used in different ways by different biologists, and our definitions do not correspond exactly with the majority usage within technical biological literature, but they are clear enough for our purposes here.

6. Darwin, in *The Origin of Species* (especially in first two chapters), constantly pushes the idea that species are hard to classify, that there are many in-between forms and the differences between varieties and species are vague.

micro- and macroevolution. It only shows that the limits of microevolution may include additional forms of life or exclude them from one group, depending on the classification. The question here is not whether evolution has limits or not but rather which biological classification is better. But even assuming the simultaneous existence of different biological classifications, it does not follow that microevolutionary change is unlimited and therefore can create macroevolutionary change. If the very distinction between the two types of evolution was impossible, then evolution in biology would boil down to nothing but change over time. And the evidence for the change of species over time does not account for evidence for the origin of entirely new species. In other words, even if the distinction between micro- and macroevolution were elusive (as skeptics claim), extrapolation of evidence from one type of biological change to another type would remain just as unjustifiable as it is in the case of extrapolation from micro- to macroevolution. It simply does not matter how we call the changes or what kind of changes they are. What matters is whether or not we have evidence for the kind of biological change that is postulated in Darwin's theory.

Finally, we need to ask, how precise must our classification of organisms be in order to justify the distinction between micro- and macroevolution? Critics may present a number of examples that seem to violate the distinction, but these examples are just a few cases out of millions that follow the rule. Even the critics would agree that there is a difference between, let's say, a dog and a wolf, and that this difference is of another kind compared to the difference between, for example, a dog and an ape. If these two differences were generated by evolution, then the first type of difference would be attributed to microevolution and the second would be an example of macroevolution. Since this example does not seem to be controversial, in this case, a discussion about the limits of evolution is possible. And we could add thousands of other examples in which the division between the two types of evolution is not controversial. The argument from extreme cases (even if they are indeed insoluble) applies only to those cases, and only in these few cases the distinction may be questioned. We see, therefore, that biological classification does not need to be strict and exhaustive to make the distinction between micro- and macroevolution possible. Thus, being aware of the criticism based on borderline cases, we will nevertheless adhere to this distinction. Hence, the controversy that we address in this book concerns *biological macroevolution*, that is, evolution occurring within the biological realm and going beyond microevolutionary changes.

## c. Biological Macroevolution

Biological microevolution does not seem to be a controversial idea in contemporary debates. Even extreme creationists do not have problems with accepting facts such as bacteria acquiring resistance to antibiotics or foxes adapting to a polar, desert, or steppe environment. Some of these facts were known for centuries, and many have been discovered recently and many have been explained in greater detail by contemporary biology. What causes the controversy is the idea of *biological macroevolution*, consisting of three major claims:

1. *Universal common descent*, meaning that all living and extinct beings descended from the first primitive organism. There is a physical connection between all living beings through natural generation.

2. *Transformation of species*, meaning that one species (natural species, as defined above) may transform into another over a long period of time by means of generation, variation, and selection.

3. *The natural emergence of new species*, meaning that no divine action transcending the powers of nature (whether mediated or direct) is required for new species (natural species, as defined above) to emerge (for more about this, see III,3).

As we can see, biological macroevolution is the idea that all life has one common origin and that the emergence of all biological novelty is a completely natural process, even if supervised by divine providence in the same way as any other natural event. At the same time, the concept of biological macroevolution does not include any particular evolutionary mechanism. In contemporary biology, there are several proposed mechanisms of evolution; however, most of them accept the basic Darwinian ideas.[7]

---

7. Stephen C. Meyer presents a few mechanisms currently proposed as alternatives to the neo-Darwinian mechanism. Meyer explains why these new attempts to save biological macroevolution do not overcome the difficulties of neo-Darwinism. See Meyer, *Darwin's Doubt*, 291–335. M. Ryland singles out six mechanisms of evolution and claims that all of them have the basic idea of random variation and natural selection as their core engine of change. See Ryland, "What Is Intelligent Design Theory?," 46–57, 48. Michael Behe speaks about "neutral theory," "multiverse theory," "complexity theory," "self-organization theory," "evo-devo," and a few other proposals alternative

For Darwin, the mechanism of evolution consisted of random variation and natural selection. For neo-Darwinism, it is natural selection working on random genetic mutations. A French naturalist, Jean-Baptiste Lamarck, who wrote a few decades before Darwin, did not really present any specific mechanism, yet he saw the cause of new species in environmental requirements and a "desire" of organisms to acquire new functions and organs. The neo-Lamarckian approach adopts the core of Lamarckism, saying that new organs and functions emerge in response to the new "needs" and that acquired traits are passed on posterity.

None of the various mechanisms of evolution proposed since Darwin is necessarily implied by the concept of biological macroevolution *per se*. The essential part of biological macroevolution includes the *effects* (not the causes) of the supposed macroevolutionary changes in the form of entirely new species that descend from the "last universal common ancestor" (LUCA). Though the particular evolutionary mechanisms are the subject of natural science (biology), the three major claims of biological macroevolution listed above belong to all levels of knowledge—science, philosophy, and theology. For example, the question of whether or not all living beings share one biological ancestor can be asked by a theologian, a philosopher, or a scientist. Each would have his or her own method of addressing this question. Nevertheless, if we adopt the Christian understanding of the relation between different levels of knowledge (as presented in Diagram 1), the answers developed at each of the levels should not contradict each other.

Contrary to what skeptics say, it is possible to compare, contrast, and even juxtapose scientific, philosophical, and theological knowledge regarding the origin of species. There are, however, two conditions that must be satisfied to make the interdisciplinary dialogue fruitful. First, we need to work out universal concepts of "species" and "evolution" that describe the same reality for each level of knowledge. Second, these concepts must be defined separately for each level of knowledge so that they can be understood by each discipline on its own terms. What makes the concepts communicable between different levels of knowledge is one common reality, described by each discipline from its own perspective. The concepts of *natural species* and *biological macroevolution* meet both criteria and thus make the interdisciplinary discussion possible.

---

to strict neo-Darwinism. However, Behe also shows that they cannot explain the origin of new functional genes and proteins. See Behe, *Darwin Devolves*, 93–137.

## 2. The Paradoxes of Darwinism

Having presented the basic definitions required by the debate over evolution we are now moving on to present some problems of biological macroevolution coming from logic (and later, from metaphysics and science). The logical problems in our context can be defined as paradoxes.

There are several definitions of the word "paradox" and, *nomen omen*, they may contradict each other.[8] Let's therefore take a definition that describes our intended use. According to an online Oxford lexicon, a paradox is "a statement or proposition which, despite sound (or apparently sound) reasoning from acceptable premises, leads to a conclusion that seems logically unacceptable or self-contradictory."[9] Thus, a paradox is a statement that ends in a logical contradiction, but it is not immediately obvious whether or how the premises are logically flawed. For this reason, unlike outright logical errors, paradoxes oftentimes live long lives as valid and coherent arguments. In this section we will see that there are serious logical problems with the principles of the Darwinian theory of evolution; however, they are not obvious at first blush. Before we get to Darwin, we will look at the phenomenon of paradoxes in a historical perspective.

### a. Paradoxes of the Ancients

Some years ago, during my basic studies at the college of my Dominican province, I was presented with one of the neatest and most entertaining philosophy texts I have ever read. It was *Paradoxes of the Ancients* by an accomplished representative of the Lwow–Warsaw school of logic (1918–1939), Kazimierz Ajdukiewicz.[10] In an engaging form, with rigorous logic, Ajdukiewicz presents some of the famous paradoxes of the ancient Greeks along with their solutions. For instance, a certain Eubulides claimed that you can never state that anybody is bald. (By bald he did not mean not having a single hair, but as we use the term casually, i.e., referring to someone who generally has no hair.) Take one man, who undoubtedly is not bald, having, let's say, 100,000 hairs on the head. Then take another man

---

8. For example, a Cambridge and an Oxford online dictionary provide contradictory definitions. Cf. https://dictionary.cambridge.org/dictionary/english/paradox and https://www.lexico.com/en/definition/paradox.

9. See online Oxford Lexicon at https://www.lexico.com/en/definition/paradox.

10. Ajdukiewicz, "Paradoksy Starozytnych."

who has just one fewer hair than the first one. Surely, this second man is not bald either, as he has just one fewer hair. Then take another man, who has just one fewer hair than the previous one. He also is not bald. If you line up 100,001 men, the last should not have a single hair, but he also is not bald. This is so, because if there were bald people in the row, then there would need to be someone who is the first bald man. But this first bald man would have just one hair fewer than the previous one, who is not bald. No one would claim, though, that a difference of just one hair constitutes a transition from non-bald to bald. Since there cannot be a first bald person in the row, there is no bald person at all, not even the last one who has not a single hair. Therefore, bald people do not exist.

Everything in this reasoning seems perfectly logical. What strikes us is just the conclusion. Yet, as Ajdukiewicz warns, you'd better not question Eubulides. Otherwise, using the same reasoning, he can easily demonstrate that you are bald and no shampoo would help you! Using Eubulides's type of reasoning, one can demonstrate that there are no adult people (only babies) or that cities (understood as settlements with a great number of citizens) do not exist. Ajdukiewicz's paper is interesting and funny, though it also leaves the reader with deeper logical considerations, and specifically, with the question: How is it possible that using perfectly flawless logic one may reach such blatantly false conclusions?

As Ajdukiewicz explains, the paradox of the bald man adopts a vague understanding of the word "bald" and then manipulates its meaning while conducting the demonstration. When the word refers to men in general a general meaning of bald is adopted, i.e., a man needs to have significantly less hair to be bald than a hairy man. But when it refers to two neighboring men, a strict meaning of bald is adopted (one hair more or less does not make a person non-bald or bald). Then the reasoning continues with the strict meaning, whereas the strict understanding of bald can apply only to two neighboring men, not to the whole row of men. Thus, the strict understanding of bald is extrapolated beyond where it is applicable. We see, therefore, that the reason for obtaining such an untrue conclusion is the free use of one term in two meanings. The logic of the argument is correct, but the meaning of "bald" changed during the argument from "generally not having hair" to "one hair less does not make one bald." Indeed, by cleverly manipulating meanings of terms, one can prove whatever one wants, and the opposite too.

A few years after reading Ajdukiewicz's paper, I became familiar with Darwin's *The Origin of Species*. After reading the first few chapters, I realized that there was a striking similarity between Darwin's arguments and those presented by Eubulides or Zeno of Elea. To be sure, they spoke about completely different realities. The similarity is not in the content of the arguments, but in the underlying logic, and the way in which they employ crucial terms. In what follows, we will investigate Darwinian theory in greater detail to show how it leads to unexpected paradoxes. Later (II,3), we will identify the reason why Darwinian arguments necessarily entail these difficulties.

### b. Species Exist and They Do Not Exist

Naturalists who rejected species fixism before Darwin did not advocate universal common ancestry. For them, the controversy was mainly about the amount of change that species could undergo. If "species" was defined broadly enough, changes remained within species. For example, Georges Louis Leclerc (Comte de Buffon) could claim (against Linnaeus) that genera can change into other genera over time, and thus he rejected species fixism. But if, hypothetically, Linnaeus had set the notion of species at the level of family, let alone phylum or kingdom, then even Buffon would have admitted that species do not change into new species because these new variants remain within the broader categories. It was different with Darwin. The British naturalist was not concerned with any particular idea of species. His argument implies that the notion of species refers to any biological reality and that this reality may change to an unlimited degree. In the second chapter of his book, he says that varieties are actually indistinguishable from species, which are indistinguishable from genera. "No line of demarcation can be drawn between species," claims Darwin.[11] Now, a critical reader could ask: Why would Darwin impose such a strong thesis that clearly contradicts the experience of nature among both scientists and regular people?

The reason why Darwin has to impose his view of nature is his attempt to introduce the idea of universal common ancestry. If species were unchangeable at any level of biological classification, then universal

---

11. Darwin, *On the Origin of Species*, 485. "There is no infallible criterion by which to distinguish species and well-marked varieties" (p. 57). "No one can draw any clear distinction between individual differences and slight varieties; or between more plainly marked varieties and subspecies, and species" (p. 470). All following quotations from this work come from the first edition, though we henceforth use the shortened title *The Origin of Species* which did not appear until the sixth edition of 1872.

common ancestry would be impossible. This is a logical problem: the logical categorizing of things excludes their confusion as long as the categories refer to something real. If species were just categories imposed on nature by our thinking, they would not exist in reality. But in the fourth chapter of *The Origin of Species*, Darwin claims that random variation and natural selection strengthen the traits of the new varieties that turn into species and eventually into new genera. This means that natural selection works in reality, not in our minds alone, so the species-specific characteristics must be produced and exist in reality. Darwin put forward his theory precisely to explain the actual existence of species, not just common characteristics of individuals. Yet, if species are entirely fluid, they cannot exist in reality. On the other hand, if they are produced by natural selection, they need to exist in reality, because natural selection, on Darwin's account, produces species. Hence, this Darwinian reasoning ends with the conclusion that species do not exist (in reality) while they do exist (in reality).

The proponents of Darwinian theory would probably respond that in Darwin's terms species do exist and do not exist in different aspects. But what are those different aspects? Darwin does not define them. And if we look at his reasoning carefully, we realize that those different aspects are impossible to define within his conceptual framework. In the second chapter he claims, for example, that fertility—the commonly recognized criterion of biological species—is completely relative. This leads to an odd conclusion that we cannot say which individuals can and which cannot interbreed.[12] Since fertility (the ability to propagate) is one of the most distinctive species-specific features, according to Darwinian reasoning, natural selection—producing and strengthening species-specific features—would need to create fertility within the species and infertility between species. Therefore, the distinction of species based on fertility is denied in the beginning of the argument, only to be reintroduced at the end of the argument. In Darwin's argument, fertility turns out completely relative; it does not exist as the criterion for distinguishing species. Yet, it is established and strengthened by natural selection in the process of

---

12. "It is certain . . . that for all practical purposes it is most difficult to say where perfect fertility ends and sterility begins" (*The Origin of Species*, 249). "It can thus be shown that neither sterility nor fertility affords any clear distinction between species and varieties; but that the evidence from this source graduates away, and is doubtful in the same degree as is the evidence derived from other constitutional and structural differences" (249).

species production. Apparently, the limits of fertility exist and do not exist. And the same problem applies to any other feature or trait that could be the criterion for distinguishing species: first, Darwin needs to deny the existence of a species-specific feature in order to introduce the transformation of species; then, he confirms that his mechanism generates this very feature in order to make new species. Thus, his argument ends with a conclusion that species-specific features exist and do not exist at the same time.

Darwin's defenders could say that the species-specific features do not exist at first but start to exist later, when natural selection generates them. In this way, a particular feature does not exist and exists, but not at the same time. Rather it begins to exist. This, however, is contrary to the logic of the argument. If this were the case, Darwin could not claim that species-specific features (such as fertility) do not exist *now*, since they have already been generated by natural selection. Yet, Darwin does claim that species-specific features do not exist *now*. For example, for him, fertility is not the criterion of distinction between the species that exist *now*. Accordingly, fertility does not exist as a species-specific feature for this particular organism as it is now; yet, it has been generated as a species-specific feature for this particular organism as it is now. Therefore, fertility (like any other species-specific feature) exists and does not exist simultaneously. Incorporating a temporal aspect does not help to overcome the difficulty.

It is true that Darwin's reasoning proceeds from one premise to another. Indeed, there is a *logical* sequence in his argument, but the *logical* sequence does not make up for the *real* sequence of events in time. As a result, Darwin needs to assume the nonexistence of species in reality in order to justify his theory of the origin of species. The theory, however, is to explain the real existence of species, which cannot be done owing to the initial assumption that species do not exist in reality. Thus, there is no way in Darwin's theory to establish which species-specific features exist (regardless of species nonexistence) and which are changeable and relative (regardless of natural selection establishing new species). This problem is embedded in the very logic of the theory. If the contrary were true, that is, if we could differentiate in Darwinian terms the permanent from the changeable aspects of living beings, then species could not evolve into other species—they would be established by the stable species-specific features. In this case, however, natural selection would not produce those features, and, consequently, not everything in biology would be produced by evolution, which is contrary to Darwin's conclusion. Therefore, Darwin's theory ends up in a paradox that species exist and do not exist in the same respect.[13]

### c. Natural Selection Simultaneously Builds Up and Ruins Species

The paradox of simultaneous species existence and nonexistence is apparent when we approach Darwin's theory statically. However, Darwin and his followers often encourage us to see his theory dynamically, as an ongoing process constantly challenging all biological reality. In order to justify the "dynamic approach" to nature, Darwin readily quoted the popular saying *natura non facit saltum* ("nature does not jump").[14] For Darwin, nature cannot jump because natural selection requires smooth transitions between each stage of evolutionary development. If nature were a smooth, non-jumping process of change, as Darwin wanted to see it, then the logical problem of simultaneous nonexistence and existence of species would be thoroughly concealed under the idea of evolution.

---

13. One more attempt to save the Darwinian logic would be to say that species-specific features become stable only after they are established by natural selection as species-specific features. Thus, every organism has a mixture of species-specific features and features that are not species-specific but would become species-specific if natural selection continued its work. This argument, however, is nothing but a mixture of the two problems already described. By the very logic of the definitions, any feature that is not species-specific cannot account for the distinction of species. The point of interest is not the origin of any features, but of those which constitute species. Darwin tries to explain the origin of species, and by this he means the origin of all biodiversity. Therefore, even if the species-specific features are generated via non-species-specific features, the problem of generating the species-specific features remains unresolved.

14. In the medieval Christian thinking, the universe was ordered but not continuous in the Darwinian sense. It was open to divine interventions. The variety of beings was referred to as *scala naturae*, which consisted of ascending grades of living beings from the most primitive to the most advanced with man on the top (in the visible order). In modernity the concept of *scala naturae* was replaced with *natura non facit saltum*. For example, to Newton it meant that nature, if it were comprehensible, could not behave capriciously, and that we should begin our inquiries with the assumption that the universe is simple until proven complex by empirical means. To Leibniz it meant that changes in nature necessarily occur gradually, that no "gaps" interrupt nature's intrinsically continuous gradations. Even Lyell's uniformitarianism derived from the same principle of *natura non facit saltum*. Lyell applied it to geological processes only; he explicitly denied the possibility of transformation of species. Darwin, however, expanded the principle of *natura non facit saltum* to living beings. On the history of the development of *natura non facit saltum*, see Woodrow Denham's (unpublished) paper, "Two Millennia of Natura Non Facit Saltum Is Enough."

Since species incessantly emerge in an ongoing process, they exist and do not exist at any time—they are in the process of becoming. One can freely claim either existence or nonexistence of species depending on which part of Darwinian theory one currently wants to justify. The very notion of species is dissolved in a more general idea of an ever-evolving biological reality. Yet, if something constantly changes, how can we say anything true about it, besides that it changes? If nature were just a process of change, as Darwin seems to say, distinctions between species would be always relative to one moment and one population or individual. But in fact, we do distinguish species, so there must be something permanent and objective in nature that enables us to do so.

This understanding of nature as a constant process of ever-growing complexity (which, we will argue, is not supported by the facts of nature) encounters another problem posed by a commonsense logic. If we take the struggle for life and the survival of the fittest as two factors that drive natural selection, it looks as if the process that supposedly creates species is the same process that ruins them. According to Darwin, natural selection produces traits for new varieties that are later fixed in the populations in such a way that the varieties are transformed into new species, species to genera, genera to families, etc. The generation of new phyla and kingdoms requires further production and strengthening of species-specific features. From the theoretically unlimited variety of possible evolutionary paths, natural selection needs to *choose* those that will lead to distinct species.[15] Next, within a given species (genus, family, etc.), natural selection needs to collapse a virtually infinite number of possible individual features into one coherent set that constitutes a given species (genus, family, etc.). Thus, in a sense, natural selection needs to arrange and classify organisms, make them distinct from other organisms, and define them according to their own species (genus, family, etc.). This is what natural selection should constantly do. Yet, at the same time, the same process of natural selection should make organisms stray from their original type in order to generate new varieties that will become

---

15. Michael Behe points out one more problem with the idea of natural selection that becomes apparent in our discussion. Behe says that in the metaphor of natural selection the idea of conscious choice is hidden: "[F]rom Darwin's day onward many biologists attributed to natural selection the ability to finely sift mutations so that, somehow, over multiple steps a coherent basis for building an organism of complex biological feature would result . . . Yet, the metaphor of natural selection trades on an equivocation. The myriad processes involved in living and dying do not *choose* anything—they just happen." Behe, *Darwin Devolves*, 202.

new species (genera, families, etc.). Darwin clearly attributes these two *opposite* effects to one process of natural selection.[16] And this is an apparent contradiction—the process of natural selection (working on random variations) simultaneously produces and ruins the established species (genera, families, etc.).

### d. The Paradox of the Necessarily Growing Divergence between the Discovery and the Theory

According to Darwin, what we have in nature today is just a tiny fraction of what existed in the past. This is supported by what we know from fossils—there were many species in the past that do not exist now, owing to extinctions that occurred over millions of years. It is estimated that over 99% of species are gone. The fossil record reveals the previous presence of this multitude of forms. Nevertheless, if we adopt Darwin's view of nature, the number of forms that must have inhabited the earth is far larger than what the actual fossil record represents.

To understand the problem, we need to distinguish between a species and an individual. When paleontologists find a specimen of a fossilized individual, they can either classify it to a species that is already known, or—if it does not fit the existing classification—they establish a new species. Thus, for paleontologists it is enough to find a single specimen in order to know what kind of species existed. By finding more fossilized specimens of the same species they may approximate where

---

16. "I attribute the passage of a variety, from a state in which it differs very slightly from its parent to one in which it *differs* more, to the action of natural selection" (*The Origin of Species*, 53). "Natural selection *acts solely through the preservation* of variations in some way advantageous, which consequently endure" (110). "*New species* in the course of time are formed through natural selection" (111). "Only those variations which are in some way profitable *will be preserved* or naturally selected . . . ; for this will generally *lead to the most different or divergent variations . . . being preserved* and accumulated by natural selection" (118). "The *divergent* tendency of natural selection" (124). "This *principle of preservation*, I have called, for the sake of brevity, Natural Selection" (128). "Natural selection, also, leads to *divergence* of character" (128–29). "Natural selection will accumulate all profitable variations, however slight, until they become plainly developed" (135). "I can see no reason to doubt that natural selection will continually *tend to preserve* those individuals which are born with constitutions best adapted to their native countries" (143). "Natural selection *acts solely by the preservation* of profitable modifications" (173). "In living bodies, variation will *cause the slight alterations*, generation will multiply them almost infinitely, and natural selection will pick out with unerring skill each improvement" (190). (Emphases added).

and when the species started and ceased to exist. But this does not tell us much about the relation of one species to another. We simply have different species in different times and places. According to Darwin, all species are connected by biological generation, so that there is a physical continuity among all individuals and—by extension—all species. Therefore, in order to corroborate Darwin's theory, it is not enough to find an individual representing a species. It is necessary to find all the links in between. According to Darwin, these links are not "true" species; they are transitory forms. But transitory forms are either "incipient" species or just individuals. If they are "incipient" species there must be still additional forms that link the "incipient" species with other "incipient" species. Finding the incipient species would still leave us with some leaps in nature. Since, according to Darwin, "nature does not jump," a fossil record required by Darwin's theory boils down to finding most or all individuals that connect two species.

Here we encounter a problem: the more new species are found, the more intermediate forms are missing, and—owing to the fact that there should be more intermediate forms than "distinct" species—the number of missing links is growing much faster than the number of species discovered. In fact, every newly discovered species creates a gap between itself and other species that requires filling with a number of individuals or "intermediate" species in order to fit the theory. Hence the problem is that the number of newly established species may be growing in a linear way, but the number of missing links grows exponentially, never catching up with the actual fossil record. Consequently, the gap between the theory and the data also grows exponentially, and the more discovery, the less confirmation for the theory.

For Darwin the lack of fossil evidence did not testify against the theory, but against the evidence. Darwin maintained that the links were missing in the fossil record because they were not preserved. Darwin rejected another option, namely, that the links were missing because they never existed. He hoped that over time the missing links would be unearthed. But the problem described here points in an opposite direction. The more new forms are classified, the more data is missing to corroborate the theory.

### e. The Paradox of Undefined Species

Darwin's argument encounters an even more general problem that he never resolved. It is a problem of defining the amount of difference between two supposed evolutionary links. According to Darwin, we should be finding millions of fossilized links because this is what life actually looked like—every species was once connected with a previous species via many transitional forms. But how many links should there be in transitioning from one species to another?

Let's imagine that we found one complete evolutionary path (though many scholars doubt that a single complete evolutionary path has ever been discovered). We have at hand all of many intermediate forms, perhaps thousands, if not millions, of them. We see that they actually change in one direction, from the initial species toward the final species. All forms are arranged exactly according to the Darwinian prediction, and they appear in the proper chronological order in the geological strata. If a Darwin-proponent had anything like this at hand, he would probably be able to convince many Darwin-skeptics of biological macroevolution.

Perhaps he could, unless the skeptics were to reflect more and come up with new questions. For instance, how can we know that our chain of smooth transitional forms is actually complete? How can we know that between each of the neighboring forms there should not be another that would fill the gap? The chain is deemed complete by the Darwin-proponent, but how can it be proved that the chain is actually complete? The Darwin-skeptic could justly claim that the evolutionary chain is far from complete, because the gaps between each of the presented forms should be filled in with additional links. Moreover, says the skeptic, if you have just three transitional forms between two distinct species, you need only four additional fossils to fill the gaps in each transition. But if you have a million links, then you need another million plus one to fill all the remaining gaps. And even if you do this—continues the skeptic—I will ask you to fill in the remaining gaps again and again ad infinitum. And this cannot be done.

Of course, the Darwin-proponent does not want to perform a never-ending task. To fend off the challenge, he needs to define the longest possible evolutionary step (i.e., the greatest possible change in one generation) and then show that none of the transitions between the neighboring forms is greater than this. But how could it be measured? According to neo-Darwinism, the longest possible step would equal the greatest amount of change that does not destroy the organism and at the same time can be "seen" by natural selection (i.e., the greatest change that is "beneficial" and hereditary). However, experiments consistently show that the neo-Darwinian mechanism can actually do very little (more about this in II,5,f) and the amount of change that random mutations can produce is not big enough to produce new functions (new functional proteins).

Outside of the neo-Darwinian mechanism, it is impossible to observe and define the longest possible evolutionary step, for at least two reasons. First, we are unable to perfectly measure the total amount of change between parents and offspring in one generation. (We can measure, for example, the amount of genetic mutation, but this is not equivalent to measuring the total change of the whole organism). And this inability is not due simply to the current state of science. It is a theoretical obstacle that can never be overcome, owing to the impossibility of accounting for every particle and sub-particle in any material being. Second, even if we measured the exact amount of change in one generation, we could never prove that this corresponds to the amount of change between the two fossils that appear to be two neighboring evolutionary links. Very often we are unable to extract DNA from the older fossils, and there are other than DNA sources of information, such as epigenetic. Some of this information is lost with the death of an organism. Hence, it is impossible to define the largest evolutionary step, and accordingly we cannot say how many links are missing (if any) in a supposedly complete evolutionary chain.

Moreover, if the Darwin-skeptic were not particularly kind to the Darwin-proponent, he would inquire further: If we cannot know whether or not the amount of linking forms found in the supposedly complete evolutionary chain is satisfactory, how can we know that the number of forms found in the supposedly incomplete evolutionary chain is *not* satisfactory? Since any supposedly complete chain may be incomplete, we can also say that any supposedly incomplete chain might actually be complete. We cannot know this unless we define the possible and actual evolutionary steps and compare them with fossil evidence, which, as we said, is impossible.

If this were not enough, the Darwin-skeptic might raise yet another problem, this time pertaining not as much to the linking forms as to what Darwin calls "true," "parental," or "distinct" species.[17] According to Darwin,

17. Darwin uses also a few other words, such as "aboriginal," "aboriginally distinct," "independent," "dominant," "well-marked," "well-defined," and "original species." The

# Evolution and Natural Knowledge

evolutionary chains (whether supposedly complete or supposedly incomplete) take us from species A to species B through countless linking forms. We can visualize this process as depicted in the following diagram:

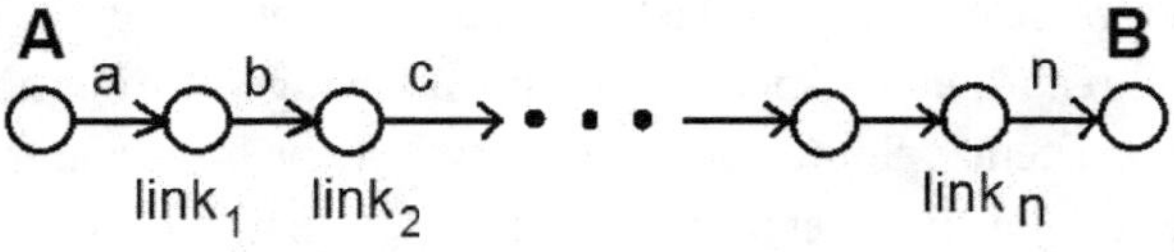

Diagram 9A. Evolution from species A to species B.

Circles A and B symbolize the "true" (parental/distinct) species. The distances between generations (a, b, c, ... n) symbolize the amount of biological change in one evolutionary step, which, as we said, cannot be clearly determined. Now, since we do not know the length of the distances between the links (distances b, c, ... [n−1]), neither can we know whether or not these distances are greater than the distance between the links and the "true" species (distance a between A and $link_1$, and distance n between $link_n$ and B). How can we know, therefore, that species A and B are the "true" species, and that the linking forms ($[link_1]$, $[link_2]$, ... $[link_n]$) are not the true species? Since the links may differ from each other as much as the species may differ from the links, there is no criterion in Darwin's theory to establish anything like "true" species as distinct from merely transitional forms. Thus, the very idea of species is nullified. In fact, the notion of species boils down to the notion of an evolutionary link which, in turn, is nothing other than an individual. Hence, in Darwinian theory, species *do not exist*; what exist are many individuals that differ more or less from one another.

The Darwin-proponent could help the theory by claiming that "true" species are either groups of individuals that are "more distinct" from transitional forms or "last longer," or both. This was pretty much the way in which Darwin defended the idea of "true" species. However, he does not offer any criterion for determining what it means to be "more distinct." In fact, if we look at Darwin's scenario, any species is

---

linking forms he calls "doubtful species," "incipient species," "sub-species," or "intermediate species."

more distinct from a more distant evolutionary form and less distinct from a closer evolutionary form. There is no reason to believe that [link$_1$] differs from [link$_2$] more than [link$_1$] differs from species A. Again, from this perspective, any transitional form could be a distinct species or just a transitional form, depending on which other evolutionary form it is compared to. The criterion of "lasting longer" is not less illusory, because in the fossil record every species remains unchanged throughout all the time of its existence. Some species simply last for longer and some for shorter periods of time before they become extinct. And the extinction of a species is not a beginning of another species, but merely the cessation of the existence of a species.

Finally, since none of the so-called "true" species are actually permanent (rather, they continue evolving along with other species), any evolutionary chain is just a part of another chain. This is depicted in the following diagram (Diagram 9B).

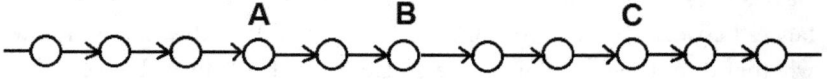

Diagram 9B. Evolutionary chain of linking forms.

If this is the case, how can we know that any evolutionary chain actually takes us from A to B or from A to C and not a few or many links too far from or too close to the starting point? The assignment of "true" species in this chain is completely arbitrary. Moreover, if we strictly adhere to Darwinian assumptions, there is not even a way to select and define an evolutionary path, because we do not know where to begin and where to end our selection. Whichever path we choose to show the emergence of a new species, it either takes us to a new species or to a linking form, and we cannot determine which of these is actually the case. Hence, none of the chains represents the transition between species because we do not know when a species begins or when it ends.

To summarize the discussion with the Darwin-proponent, the Darwin-skeptic would present three unresolved paradoxes:

# Evolution and Natural Knowledge 109

1. One link in an evolutionary chain may not be enough to justify the transition between species, but we do not know if a million links are enough. At the same time, one link could be enough and a million links could be superfluous. But it may also be the case that any number of links is not enough, and then one would need to infinitely look for the links, which is impossible to accomplish.

2. Since the transitions between species must be smooth, according to Darwin, the evolutionary steps are so small that they are hardly visible. However, if they must be small, they cannot be large in the case of transitions between the links and the "true" species (a and n in Diagram 9A). Since no transition differs essentially from any other transition, we cannot tell where the "true" species begins. Therefore, transitional species are the same as distinct species because distinct species do not differ more greatly from the transitional species than the transitional species differ from each other. And this boils down to the non-existence of species. What exists are just individuals that differ more or less from one another. But if species do not exist, one cannot explain their origin.

3. To demonstrate Darwinian theory, one may present an evolutionary chain of transition from one species to another. However, because we do not know where the distinct species are, we also do not know where (if at all) any evolutionary chain begins or ends. Thus, within the Darwinian theoretical model it is impossible to present even one evolutionary chain. Only the totality of all living beings constitutes a single chain that begins with the first living organism and ends with the last living organism. Yet, this one great chain does not show the origin of some or all species but only the origin of one species, namely, the one at the end of the chain. And even this is not certain, because also the last one constantly evolves.

The reason why Darwinian theory sinks in paradoxes is the lack of clear definitions of terms such as *distinct (true, parental) species* and *intermediate (doubtful, incipient) species* (i.e., links). However, as we said above, blurring these notions is the initial condition for pursuing the theory. Darwin could not adopt a clear definition of species, because he would have needed to distinguish between something changeable in the biological realm and something permanent. A well-defined "permanent thing" could not be explained by the process of change, and thus it would remain unexplained within the theory. Darwin's theory simply cannot be presented in clear terms. This is why any additional definitions and explanations of species (employed, for example, by later Darwin-proponents) work well only in certain limited contexts when applied to some aspects of the theory—not within the entirety of the theory. The confusion and lack of clear definitions is indispensable for Darwin's thinking. His reasoning cannot be clarified, and if it were, his theory wouldn't stand. We can conclude that Darwin's theory feeds on confusion, while clarity makes it illogical.

### f. The Paradox of Random Variation and Natural Selection

According to Darwin, new species arise through the production of new characteristics and features in old species. Among the new features, the most distinctive and significant are new organs. This is why Darwin, in order to explain the origin of species, had to explain the origin of new functional organs. Darwin saw that an organ performing the same function may be more complex in one species than in another. This made him believe that the organ evolved from the primitive form found in one species to the biologically advanced form through a process of gradual improvement.

Yet, by discussing the evolution of an organ, Darwin did nothing to explain how the organ originated in the first place. In fact, the origin of organs for Darwin remained an unresolved mystery. We can see it when he speaks about his best example—the eye. He readily selects a variety of eyes found in nature and orders them according to growing complexity. Still, he admits that it is difficult to accept that the eye was produced by a blind process of natural selection: "To suppose that the eye ... could have been formed by natural selection, seems, I freely confess, absurd in the highest possible degree."[18] Thus, if we are to accept the origin of the eye through natural selection, "reason ought to conquer imagination."[19] Finally, when Darwin reaches the point of explaining the origin of the *first* eye, he can hardly hide his irritation: "How a nerve comes to be sensitive

---

18. *The Origin of Species*, 186.

19. *The Origin of Species*, 188. As Michael Behe points out, this was a clever rhetorical trick—Darwin had actually reversed the roles of reason and imagination. (Behe, *Darwin Devolves*, 47.) In fact, it is imagination that has to overcome reason. Only after abandoning reason and escaping into the realm of imagination could one believe that a complex organ such as an eye would simply "come to being" without any intelligent design.

to light, hardly concerns us more than how life itself first originated."[20] At the climax of the argument, Darwin leaves the reader with the impression that his theory does not explain how the first life-form, or the first organs, originated. But if it does not explain these two things, it hardly explains anything relevant.

To save the theory, the followers of Darwin respond that the problem of the origin of new organs has been resolved by the process called exaptation (also known as co-option). According to this idea, new organs are simply older organs that were adapted for performing new functions. Clearly, exaptation does not explain how organs originated in the first place, but only speculates of how they may have adapted in order to perform new functions. Since exaptation implies the preexistence of functional organs, it is a question-begging solution.

The problem of the origin of new organs is connected with a more general problem concerning the Darwinian mechanism of natural selection working on random variations. The logic of the mechanism implies that each evolutionary step must bring about something that gives advantage in the struggle for life. If random variation (for example, a heritable genetic mutation) does not produce any competitive advantage, it will not be visible to natural selection and thus will not propel further evolution. Now, what are those "variations" that give any advantage over competitors in the struggle for life? If these are only small changes that give advantage in specific circumstances, such as antibiotic resistance, fur color, or thickness, they will be preserved by natural selection only under these particular circumstances, but they will not help to move evolution in one direction beyond these minor adaptations. This is because these variations do not constitute any functional novelty. They were present in the population, and natural selection merely brought them out or strengthened them, or changed the proportion of individuals in the population according to survival circumstances. But in such cases, nothing new was made by evolution. To produce biological novelty, evolution needs to produce new functions.

New functions, however, are performed by new genes, new proteins and new organs that must be complete to be functional. Half of a wing, one feather, or an eye lacking nerves would not give any survival advantage and would be *eliminated* by natural selection as an extra burden. (The same applies at the molecular level; new function requires new

---

20. *The Origin of Species*, 187.

genes or proteins.) Thus, either a new function or an organ is complete and ready to function in one evolutionary step (one generation), or it is invisible to natural selection.

This leads us to a paradox—small changes (those that do not produce new functions) are not visible to natural selection, and large beneficial changes (those producing new functions) cannot be achieved by random variation. The paradox consists of an incommensurability between random variation and natural selection. If the former provides beneficial changes, they are too small to enable natural selection to move beyond microevolution. If it provides large changes, they are detrimental and eliminated by natural selection. Either way, the evolutionary process will only oscillate back and forth around minor adaptations and never create entirely new forms of life.[21]

The solution proposed by evolutionists is that even if the Darwinian mechanism cannot produce great changes (such as entirely new organs or species) in one large step, it can accomplish this in many tiny steps.

Let's assume that natural selection wants to move from organ/function A to organ/function B. Even though this is impossible in one step, it becomes possible if between A and B there is another organ/function C. Natural selection does not need to jump directly from A to B; rather, it moves from A to C and then from C to B. Diagram 10 depicts this process.

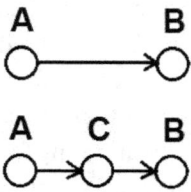

**Diagram 10. Evolution from organ/function A to organ/function B via organ/function C.**

Does this scenario resolve the problem? Apparently it does not, because the transition from A to C must be of the same nature as the

---

21. Michael Behe provides many actual examples and biological details of how natural selection fixes species in their niches, but at the same time limits their ability to adapt to new conditions. Behe calls natural selection a "self-limiting" process. See *Darwin Devolves*, 197–227.

## Evolution and Natural Knowledge

transition from A to B. The solution implies that the A–C transition is easier than the A–B transition. But how could it be easier? Both transitions need to generate biological novelty in order to be visible to natural selection. Therefore, the A–C or the C–B (or both) transitions encounter exactly the same obstacles as the direct transition from A to B.

If we strictly adhere to the logic of the Darwinian mechanism, we see that the proposed solution does not fix the problem but actually multiplies it. Now natural selection needs to transition twice: from A to C and then from C to B. We can, of course, hypothesize about additional intermediate stages, but this would again multiply the problem, not resolve it. The longer the chain of links, the more transitions there are between A and B, and the more difficult it is for the blind process of natural selection to continue in one direction. Finally, if the chain is infinite, natural selection would never cross it. Hence the paradox—the real macroevolutionary steps would be too long for random variation and natural selection, but the short attainable steps are too short to produce the required changes.

*❦*

We said above that the logic of Darwinian evolution resembles the logic of Eubulides and Zeno of Elea—the ancient authors of paradoxes that entertained the ancient patricians. They wanted to prove things such as the impossibility of physical motion or that there are no bald men. Yet, their conclusions are clearly against our experience and the facts. The task they left for the future was to find the errors in their reasoning. Scholars who resolved the ancient paradoxes were right in their conclusions that physical motion is actually possible and that some people are bald while others hairy. Someone could therefore raise an objection against the paradoxes that we found in Darwin's argumentation: If the ancients were wrong when proposing the impossibility of motion and the like, perhaps our claim about the logical impossibility of evolutionary transitions from one species to another is also mistaken?

Let's take, for example, the paradox of the bald man, mentioned above. The argument begins with a line of men. Every man in the row has just one fewer hair than his preceding neighbor. Since one fewer hair does not make a man bald, even the last man in the row, who has no hair at all, should be hairy. The paradox of Darwin's argument is analogous: Imagine that a species B evolved from a species A. Each organism belonging to the transitional chain from species A to species B differs slightly from the previous one. In subsequent generations, A gives birth to $A_1$, which gives birth to $A_2$, which gives birth to $A_3$, etc. According to Darwin, at the end of the chain there will be **B** instead of $\mathbf{A_n}$, exactly as we find a bald man at the end of the chain instead of a hairy. The charge against Darwin is that we will never get **B** but only $\mathbf{A_n}$. Hence, the transition from species **A** to species **B** is impossible.

The logical structure of both paradoxes—the ancient one and the Darwinian one—is the same. Yet, as we said, the ancients were wrong. How can we, therefore, claim that the Darwinian paradox is a legitimate charge against Darwinian theory and not just a logical trick like the paradoxes of the ancients? Perhaps our argument is based on a mistake, just as the paradox of the bald man is based on manipulating the term "bald"? To address this issue, we need to connect Darwinian theory with real being through metaphysics. Only then can we show how the Darwinian paradoxes differ from the ancient paradoxes—and thereby confirm that they are valid charges against Darwinian theory. Thus, we will respond to the present difficulty (in II,3,b) only after introducing a few metaphysical terms.

## 3. Evolution and Metaphysics

Until now we have presented some problems with the *logic* of the Darwinian idea of transformation of species. There is, however, a *metaphysical* argument testifying against the very possibility of biological macroevolution, regardless of its specific mechanism. Today, this argument hardly exists in the debates over evolution. The reason is not that the argument itself is weak or invalid but rather the fact that classical metaphysics gets little regard, understanding, or appreciation in contemporary culture. A hundred years ago Catholics still held classical philosophy in high esteem. This is why many ecclesiastical scholars combated Darwin's ideas by resorting to classical metaphysics.[22] Today, the situation is different: a majority of Catholic scholars, whether scientists, theologians, or philosophers (and particularly philosophers of nature) claim that biological macroevolution does not pose a serious difficulty for Christianity. This complete reversal, which occurred about the middle of the twentieth

---

22. For a detailed account of those debates, see my book *Catholicism and Evolution*, especially chapter 4, 85–145.

century, is by itself an interesting phenomenon which we are not going to discuss here.[23]

In our argument, we will appeal to the so-called eternal philosophy (*philosophia perennis, sana philosophia*), which consists mainly of classical metaphysics. It is a somewhat broader philosophy than that of Aristotle and Thomas Aquinas, though these two scholars are the best representatives of it. In our opinion, generally speaking, Catholic philosophers of the late nineteenth century understood the problem of evolution better than their contemporary counterparts. Even if they did not know modern biology and sometimes fell into somewhat naive biblical interpretations, their struggle with the evolutionary paradigm was inspired by an honest commitment to, and a deep understanding of, the eternal philosophy. They were well equipped to define the contradictions between classical metaphysics and biological macroevolution. The greater part of their work, however, remained hidden in the Vatican Archives or in Latin theology textbooks.

Over the last seventy years, the arguments of Catholic philosophers have been overridden by the teachings of a more vocal group of "compatibilists," that is, those who believe that classical metaphysics do not challenge the idea of biological macroevolution. Many compatibilists say that we cannot even compare theological, philosophical, and scientific ideas because of their so-called incommensurability. According to some compatibilists, different disciplines address completely different matters, and each should be kept within its own domain. Compatibilists usually do not absolutely exclude any and all dialogue between theologians and scientists, yet they exclude the very possibility that a scientific idea could be countered or disproved by theology or philosophy. Hence, many compatibilists are effectively "isolationists," in the sense that they do not see a real connection between different levels of knowledge. In our opinion, the knowledge found in different levels may, without any confusion, be compared and juxtaposed and even contrasted, as long as each level refers to the same reality and defines this reality in its own terms (see our previous discussion of the "two books" in I,1,c and the definition of natural species in II,1,a). Let's see now what kinds of obstacles classical metaphysics presents to biological macroevolution.

23. A brief elaboration of this problem, with the indication of possible causes of the shift that happened in Catholic philosophical circles, can be found in my book *Aquinas and Evolution*, 223–37. For examples of how Catholic philosophers combated Darwin's theory in the early ecclesiastical debates, see my *Catholicism and Evolution*, chapter 4.

### a. An Accidental Change Does Not Generate a Substantial Change

According to classical metaphysics, there are several divisions (or contraries) in being. One of them is the division into essence and existence. A connection of the two constitutes the existing being.[24] Since essence and existence are not the same, any created being may exist or not. In God, however, essence and existence are identical. For this reason, He alone cannot cease to exist. In a material being, the essence (or the substance) of the being is characterized by accidents. For our argument, the distinction between the essence and accidents is crucial. Essence is what a being *is*, namely its substance, whereas accidents are what a being *has*. We can also say that accidents are the features of a being, or characteristics that describe the substance. Our human cognition of a material being comes through sensual perceptions. By sensual contact with things, we capture just the accidents of a thing. We discover the essence of the thing through the operation of our mind called abstraction. The important postulate of classical metaphysics is that the essence of the thing exists in every particular being. It is not just a construct of our mind, and it does not exist only in some ideal realm, but in reality, in every individual being. This is why we discover, or derive the essence from things, rather than create it or impose it upon things.

Another division in a material being consists of form and matter. A compound of form and matter constitutes the substance of a material being. This is why material beings are called "composites," in contrast to immaterial beings, such as angels, who are "simple." In a composite being, form accounts for all specification of matter—everything that gives matter any quality, any shape or nature. We can say that form is everything that we can grasp intellectually regarding any being. The very word to *in-form* derives from the activity of our mind by which we capture intellectually the forms of things. In contrast, matter is an underlying substrate of a material being, a pure potency, something that cannot exist by itself yet makes it possible for the form to create a material being when form and matter are combined. Form makes a being intelligible, but matter allows it to become a material individual. Hence, form may also be considered the entirety of a being's intelligibility—everything that we can say in a positive way about a being. Every material being has a substantial

---

24. Just for the sake of precision: existence cannot be seen as a piece that is added to essence in order to obtain an existing being. Rather existence is an act of essence. Existence is not what a thing *has* but rather what it *does*.

form and accidental forms according to the distinction between essence and accidents.[25] The structure of a composite being is summarized visually in Diagram 11.

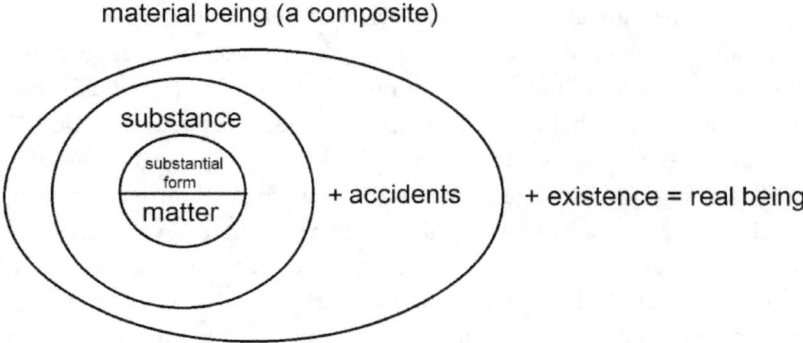

Diagram 11. Relations between substance, existence, and being.

One example of a material being is an animal, for instance, a dog. A dog has accidents, such as the color and length of its fur and the width and length of its ears and legs. Even its spatial position, temperature, and movement are accidents that characterize a dog. All of this constitutes the accidental form of a dog. Some accidents are more important than others, because they belong to every dog (the so-called proper accidents). These include four legs, one head, one tail (sometimes in a rudimentary form), one nose, etc. If we deprive a dog of any of these proper accidents, we may still have a dog, but it is not a normal or healthy dog. In contrast, the substance of a dog is the "dogness" itself, something that we cannot physically point out, but we know that it is there and that it is something distinctive for all dogs and each dog, something that differentiates dogs from other substances, such as cats or elephants. Substance considered with regard to the operation of the being is called nature. Thus, it is in the nature of a dog to act in a doglike manner. The notion of nature is close to that of substance, which is close to the notion of essence. In the argument presented here, these three may be used interchangeably.

---

25. A broader, though still simple, introduction to classical metaphysics can be found on the website entitled "Aquinas and Evolution" (www.aquinasandevolution.org).

Having introduced the basic metaphysical notions, we now need to translate the idea of biological macroevolution into metaphysical terms. This will allow us to juxtapose biological macroevolution with the principles of classical metaphysics.

Darwin's evolutionary mechanism implies that random variations occur in generation and then natural selection acts upon them in order to bring about biological novelties. Natural selection accumulates so-called beneficial variations (traits) and eliminates so-called detrimental variations (traits). But what kind of changes are those "random variations"? Whether they are genetic mutations, slight deformations, organic modifications, or any other changes generated during procreation, development, or adulthood, they are always *accidental* changes. These never change the substance of an organism. A dog remains a dog regardless of whether it is born white or black, robust or fragile, healthy or sick, etc. Therefore, no matter how many and what kind of accidental changes natural selection preserves, they can change only the accidental form of a dog, not the substantial form. As we said, the substance of an organism really exists in this organism. Hence, the substance of a dog exists in each dog, and this is what remains unchanged regardless of how many accidental changes a dog undergoes in subsequent generations. All of the accidental changes account for only microevolution that does not transform a dog into any other nature. This is why it is impossible to get a new substance, such as a cat, or a bird, or a cow, or a fish, via random variations and natural selection. Material process can change only the accidents of individuals and for this reason no material process can account for the substantial change, such as between two different natures. For this reason, biological macroevolution is philosophically impossible. Indeed, metaphysics excludes not just the Darwinian mechanism of macroevolution but any idea of transformation of species whereby physical continuity is maintained (for example by natural generation) and the only source of novelty are accidental changes—even if these changes were nonrandom but deliberately guided by a mind. Thus, classical metaphysics excludes also theistic evolution, which is nothing other than biological macroevolution somehow guided, accompanied, or initiated by God (see our definition of theistic evolution in III,1,b and III,3).

When this argument was presented elsewhere,[26] one of the anonymous reviewers of my paper raised the following objection:

---

26. Chaberek, "Thomas Aquinas and Theistic Evolution" (web article).

When I eat a banana, the banana undergoes a substantial change such that its matter, once informed by the substantial form of banana, is now informed by my soul [a substantial form of man—M. Ch.]. This process happens through a digestive process that introduces accidental changes into the initial substance that changes its predisposition to its original substantial form [i.e., banana—M. Ch.] so that it is now predisposed to my own substantial form.

The author of this objection wants to say that there are examples when an accidental change brings about the substantial change. I agree that even one such example would ruin our metaphysical argument against biological macroevolution. However, there is no valid example in the quoted objection. The importance of this objection resides in the fact that it contains not just one but two possible errors in the understanding of the argument proposed here. Let's look closer at both.

First, the objection stems from a confusion made between an individual being (an individual banana) and the species to which this individual belongs (banana). In other words, there is a confusion of the individual form and the substantial form. It is true that if we ruin a particular being it will lose its form, but this is just a form of this particular individual, not a form of all beings belonging to this species. Surely, the destruction of an individual entails the destruction of the entire being along with its substantial form; however, this refers to the substantial form existing in this individual which is an individual form. The substantial form—the one that establishes the species of banana—remains unchanged, because the destruction happens to the individual not to the species. If one eats a banana, one obviously ruins this particular banana, but there are still many bananas in the world, and each of them still carries the substance (substantial form) of a banana. Hence, the substantial form of a banana is not destroyed. And even if some angry evolutionist ate all bananas in the world (just to prove metaphysics wrong), the idea of a banana would still exist in the mind of God. Thus, he would destroy all individuals, but he would neither alter nor destroy the species, substance, or substantial form of a banana.[27] And this is contrary to what biological macroevolution postulates.

27. We find exactly this distinction in Thomas Aquinas, who teaches: "Matter is for the sake of the form, and not the form for the matter, and the distinction of things comes from their proper forms. Therefore the distinction of things is not on account of the matter; but rather, on the contrary, created matter is formless, in order that it may be accommodated to different forms" (S.Th. I,47,1, co). Similarly, he writes: "Those things whose distinction from one another is derived from their forms are not

The second and greater problem with the objection is that it is based on an imprecise understanding of what substance is. Substance is something that constitutes a unity in the highest degree. Substance is the most integral and self-contained thing; it is also the most simple and specified thing. It is something that exists "by itself," not by something else. The only true substance is God, as He is the simplest, the most specified, the most self-contained—He simply *is* in the highest degree. All created substances are substances to a different degree according to how much they participate in the substance par excellence. Substance is an analogous concept; therefore, it is applied to different things in a different way. The truest substances in the created order are the angels, because they are composed only of substantial form and existence (they have no form-matter composition). When we move down to the material universe, the lowest substances are simple compounds, such as water and air, followed by elements. They should not even be called substances because the degree of their specificity is very low and they do not constitute a unity. Instead, they are rightly called elements, because they are just conglomerations of atoms.

The banana used in the objection (a dead banana) is just a combination of different compounds, definitely higher in the hierarchy of substances than the chemical elements, but much lower than, let's say, any living plant. Plants are substances in a much higher degree because they constitute a much more specified unity than chemical elements or complex compounds, such as meat or food liquids. Above plants there are animals, which in turn are more specified, have more life and more unity; they are more actualized than plants in many aspects. The highest composite substance is man. Therefore, the truest substances in the visible order are humans and animals, followed by plants. Our

---

distinct by chance, although this is perhaps the case with things whose distinction stems from matter. Now, the distinction of species is derived from the form, and the distinction of singulars of the same species is from matter. Therefore, the distinction of things in terms of species cannot be the result of chance; but perhaps the distinction of certain individuals can be the result of chance" (*ScG* II,39,3). In another place, Thomas explains that the multitude of individuals of one species exists in order to preserve variety of species among composites: "And as the matter is on account of the form, material distinction exists for the sake of the formal distinction. Hence [. . .] in things generated and corruptible there are many individuals of one species for the preservation of the species. Whence it appears that formal distinction is of greater consequence than material" (*S.Th.* I,47,2, co). It is worth noting that in theistic evolution the multitude of individuals (populations) exist precisely for the contrary reason—to destroy existing species and bring about the new ones.

metaphysical argument applies only to "true" substances, such as plants, animals, and humans.

One way of testing "how much of a substance a thing is" can be done by asking how much of a unity a thing is. The more something is indivisible the more of a "true" substance it is, because substance is what constitutes unity in the highest degree. We cannot divide an angel at all, because it has no matter and matter is the principle of division. An animal is somewhat divisible, because we can imagine cutting off some limbs and removing some organs and still see a living animal, which means that the substantial form persists in the individual, even if it becomes a lame individual. However, disconnecting the head from the rest of the body destroys the animal substance. Plants are even more divisible than animals, which means they constitute a unity in a lesser degree. But when it comes to elements and compounds, their divisibility is almost unlimited. Water, for example, can be divided until there remains one particle of water. One particle of water probably should not be called water, which means that the substance of water was lost somewhere on the way of division.[28] It is even hard to pin down the moment when the substance of water is lost, which confirms that water is a substance only in a very weak sense. The amount of divisibility that water or any other compound can sustain is significantly higher (compared to living beings) and this confirms that compounds and elements are not substances in the strict metaphysical sense of the word (they are not "true" substances).

The correct understanding of the analogous character of substance is a principal condition for understanding our metaphysical argument against biological macroevolution. Had the author of the objection used an example of a true substance, he might have noticed the error in the objection. For instance, if one kills a chicken, one destroys the individual form and the substance of this particular chicken. Yet, the chicken does not change into any other substance, such as an eagle, a dog, or a cow (keep in mind that all of these are "true" substances only when they are alive). Hence, the accidental change never generates any new substance. It may diminish an individual substance, but not create a new one.

We see this metaphysical problem confirmed everywhere in biology, whether in nature or laboratories. When evolutionary experiments are performed to see how much change different organisms can undergo, a threshold is always encountered. For example, we can try to change a

---

28. If someone assumes that one particle of water is still water, the substance of water would be lost only after dividing the particle into elements (i.e., hydrogen and oxygen).

lizard into a snake by applying multiple mutations. But the effect is always either a lizard with so many changes, which is not a snake, or a dead lizard, which is neither a lizard nor a snake. Metaphysics explains this problem thoroughly: The changes applied to a lizard are changes in individuals which are performed on matter. A lizard can suffer change only as long as matter underlying its substantial form is proportionate to the form. If matter is altered too much ("disproportionated" from the form) the form cannot inform matter any more, and departs. The composite ceases to exist, which we see when the animal dies. We see therefore how biological experiments confirm the metaphysical problem brought up here. Truth cannot contradict truth.

We should also notice here that biological macroevolution assumes not just transformation of one substance into another but also creation of an entirely new substance, one that never existed before. Transforming one existing species (e.g., a dog) into another existing species (e.g., a cat) is obviously something different from creating an entirely new species. Therefore, a possibility of transformation of species by accidental changes would be a necessary but not a satisfactory condition for any macroevolutionary scenario. Since this necessary condition is metaphysically impossible, biological macroevolution can't happen (cf. II,3,e).

## b. Response to the Objection Raised against the Darwinian Paradoxes

Having presented the first metaphysical argument against biological macroevolution, we are ready to respond to the earlier charge against the validity of the paradoxes found in the logic of Darwinian arguments. The question was, why will the man with hair become bald if we continue subtracting a single hair from the head of each of the thousands of men in a row? And, in contrast to this, why will a reptile (for instance) not become a bird (for instance) even if we keep accumulating accidental changes endlessly in generation after generation? The reason is that in the first case (hairy turning bald) we do not deal with a change of substance but only with a change of accidents. We can also say that in the case of a man with hair, we speak about a *quantitative* difference alone, or what a man *has*, whereas in the case of transformation of species we deal with the substantial or *qualitative* difference, or what a thing *is*. The first transition is from a hairy to a bald man, which does not require a substantial change. In the second transition, however, the substance of a reptile is to be transformed into a substance of a bird. This requires the substantial change, which is beyond the reach of accidental changes. This is why the first transition is possible through the accumulation of accidental changes (like a loss of a hair), but the second is metaphysically impossible, and this is also why it has never been observed in nature. This also explains why Eubulides was wrong in arguing against the transition from hairy to bald, whereas the paradoxes found in Darwinian argumentation are valid and serious objections to the very logic of the theory of biological macroevolution.

To this we should add that most of the problems with Darwinian theory, whether in logic or metaphysics, or even in experimental science, have their source in the permanent confusion of these two things—substance and accidents. Furthermore, the very fact that rejecting the metaphysical distinction into substance and accidents leads to logical contradictions and difficulties in biological science testifies to the trueness of the distinction itself. This distinction is not just an invention of an ancient philosopher or a product of our thought; rather, it reflects the very structure of material being. It tells us "how the things are." This structure, conceived of and studied on a metaphysical level, is applicable in all other levels of knowledge, because truth cannot contradict truth.

## c. The Problem of Sufficient Cause

For decades after Darwin and Spencer proposed the idea of "creative evolution" scholars kept raising the problem of the lack of sufficient cause in the Darwinian theory. More recently, however, this crucial argument seems to be rarely quoted by the critics of evolution. There are two possible reasons why. The first is the general crisis of metaphysical thinking in our culture which makes the question about the sufficient cause look too "abstract" and "unscientific." The second is the fact that materialistic philosophers, along with theistic evolutionists, convinced the public that nature as a whole is characterized by some "hidden properties" of "self-organization" that can spontaneously generate life from nonlife and species from non-species.

There were many authors who greatly contributed to this "mental shift" in the current understanding of nature, but probably no one has had a greater impact than the French philosopher, Teilhard de Chardin.

De Chardin refers to a number of laws and properties supposedly hidden in matter to make plausible his idea of spontaneous emergence of complexity out of simplicity and order out of chaos. Teilhard refers to concepts like a "vast, universal phenomenon of complexification of matter," "molecules' capability of appearing, of germinating, anywhere, without exception, in the world of atoms," "major currents that affect the universe in its totality," "unfettered moleculization," the "glow of life," the "cosmic movement of corpusculization," "hidden laws of biogenesis," the "fundamental law of complexity-consciousness," "compressive socialization," the "latent germinal powers of the earth," and many others.[29] De Chardin also imagines the existence of some "radial" and "tangential" energies whose interaction supposedly results in the preservation of life and the emergence of higher orders of being.

It is difficult to avoid the impression that Teilhard creates these esoteric concepts *ad hoc*, whenever he needs to justify the scientifically unfounded cosmology of an emergent universe. The reality, as much as it is known by physics and biology, contrasts with the imagined universe of de Chardin. In the real world, we find very few and very poor examples of self-organization, such as the emergence of crystals. But even in these few cases, nothing really new emerges; it is only that the particles that are already present are rearranged into spatially repetitive structures. But transitions such as those between dead and living matter or vegetative and sentient life demand some other type of change and organization. In any event, even if nature can produce some complexity, it is unable to produce irreducible complexity or specified, purposeful information which is indispensable for life to exist (for the meaning of these terms, see II,5,b–e). In the light of the known facts about nature, it becomes evident that the "laws" and "powers" invented by Teilhard are pure fictions. Nothing in the real world corresponds to them. Yet, his theology-fiction enchanted the entire generation of scholars and still stays strong in many minds. How was this entire shift possible? One probable reason is the departure from metaphysical realism, along with the abandonment of Aristotelian-Thomistic strain of thought in ecclesiastical studies. Once the realistic philosophy was dropped, a philosophy-fiction in the Teilhardian style quickly flourished. As the old saying goes, *natura abhorret vacuum*.

The real model of the universe is the one showing that total available energy of the system is on the decrease. Never-diminishing entropy is a

---

29. See *Man's Place in Nature*, 19, 28, 31–32, 56, 60, 98, 109, and *The Phenomenon of Man*, 48, 77.

fact. And this applies not only to physics but to all material reality. In biology, we observe not the emergence of species but rather the convergence of races and the extinction of species. Reality stands in opposition to the theories contrived by philosophers who believe in self-organization, spontaneous generation or natural development of life into higher forms.

Biological macroevolution requires that a lower cause generate a higher effect. Many evolutionists are actually aware of this problem. In order to avoid it, some evolutionary thinkers reject the very idea of the "higher" and the "lower" in biology. They believe that in evolution there is no progress, no perfection, and no direction whatsoever. But the same thinkers maintain that at the beginning of this process is a single-celled organism, and at the end complex things such as mammals, including human beings. It is clearly inconsistent to believe that the process that transformed amoeba into man is one that has no direction, no progress, no perfection whatsoever.

Surely, we can define perfection in different ways, but for our argument it is enough to see it as a growth in complexity. This definition is understandable in modern biological terms. It is clear that a single-celled organism is simpler than a mammal—it has fewer organs, fewer functions, fewer systems, etc. Therefore, even within the materialistic evolutionary framework there is growing perfection in biological macroevolution. If this is the case, a question follows: What would be the cause of this growth?

Here we need to define this problem in metaphysical terms. In order to do it, first, we need to introduce one more division (contrary) in being. It consists of *act* and *potency*, two states of being that define being in terms of its perfectness. The most perfect is the being that is completely in act, which means that all possibilities are realized in it. And this is only God. At the other end of the spectrum is something that is pure potency, and this is prime matter (*materia prima*). It has no actualization whatsoever, because it has no form. It is completely undefined and cannot exist *per se*. Since it has no definition, we cannot even create a proper idea of it. We can know it only by understanding what it-is-not.

Between prime matter and God there are created beings that can be arranged in the chain of growing actualization. In the biological realm, plants are clearly less in act than animals. They have life (which makes them more actualized than nonliving beings), but they cannot move or sense. Animals have senses and are capable of local motion. Man, in turn, has an intellect, which makes him more "in act" than other living beings.

There are a number of smaller differences in the amount of act and potency among different species. For instance, a cat or a dog does not have a trunk, whereas an elephant has a trunk, which adds some actualization to its species (compared to a cat or a dog). The amount of act in a being determines the degree of its perfection—the more act (actualization), the more perfect the being.

Now, the problem with biological macroevolution is that no being can pass on to another being more act than it has. In order to bring something from potency to act there must be something that is already in act. Some scholars believe that this principle is suspended in ontogenesis, when a being, such as a fertilized egg, develops by itself into an adult specimen, apparently actualizing itself without an actualized agent. Metaphysics teaches, however, that any living being must have a principle of its operation that is different from matter yet completely connected with it. This is the animal soul which is the form of an animal. This immaterial form is a sufficient cause for actualizing matter during the embryonic development of an individual.

When it comes to the creation of new species, there is nothing in the theory of biological macroevolution that could actualize the potencies of matter. In generation, only as much act is passed on to posterity as there is in the parents. The accidental changes that supposedly propel natural selection are just material changes, not any higher or more actualized causes. Biological macroevolution does not explain the increase of functional information that is needed to produce new species. And this is why Thomas Aquinas says that "generation of nothing except a man results from the semen of man."[30] Analogously, nothing but a cat is generated from a cat, from dog nothing but a dog, and so forth. Hence, the concept of biological macroevolution lacks a sufficient cause. It boils down to saying that something happens by itself. New species simply pop up out of nowhere. Biological macroevolution is therefore impossible.

### d. The Problem of the Reduction of Causes

Darwin and his followers firmly defend the idea that the evolutionary process has no goal or direction. This conclusion is a consequence of adopting the Darwinian mechanism of evolution, which is based on interaction between chance and necessity. The chance factor is random genetic mutations,

---

30. Aquinas, *Summa contra Gentiles* III,69,c.

and the necessity factor is natural selection. Random mutations are unpredictable, unplanned, and not guided by anything. They result from pure chance. Natural selection, on the other hand, is a necessary process that usually preserves beneficial adaptations, does not "see" neutral variations, and eliminates detrimental variations. What really drives the evolutionary process in the Darwinian and neo-Darwinian sense is chance. Yet, Christian theology and philosophical tradition strongly support the existence of purpose and goal in the natural world. Hence, the problem with Darwinian evolution commonly recognized by Christian philosophers is its lack of teleology, or in other words, a lack of *final cause*.

To resolve the obvious conflict between (neo-)Darwinian evolution and Christianity, many Christian philosophers resort to the idea of "guided evolution." They say that the overall evolutionary process, even if random in details, is guided by divine intellect on a higher, "theological" level. God works as a first and a final cause in this process. Compatibilists (those who believe that evolution and Christianity are compatible) also believe that this idea perfectly harmonizes with the teaching of Thomas Aquinas, who says that God can bring about His intended goals using both planned and chance events.[31] In evolution, God simply uses random events to achieve His preplanned ends, such as the emergence of species. Thus, evolution is random on the empirical level but non-random in the light of faith. It serves as a secondary cause of divine creation. As such it reveals finality and, consequently, does not contradict the Christian faith. This solution is called theistic evolution. It appears so simple, neat, and convincing that it has been adopted nearly universally across Christian denominations, among both academics and laymen. Yet, the enchanting coherence of this concept thoroughly conceals the fundamental metaphysical difficulties it entails. Before we ask if this solution is metaphysically possible, we need to introduce the concept of the four causes.

According to classical metaphysics, to explain any material being in terms of causality means to detect four causes that make the being possible. If any of the four causes is missing, we cannot explain the being—indeed, the being cannot even exist. The four causes are the material, the formal, the efficient, and the final. A typical textbook example for explaining the four causes is a marble statue created by a sculptor. The *material cause* is the marble (the unshaped and undefined matter

---

31. This is the way in which the problem of evolution and Christian faith is resolved by the International Theological Commission in its 2004 document *Communion and Stewardship*, point 69. For our response to this argument, see III,3,c.

underlying the form), the *formal cause* is the actual shape of the statue (the form that specifies the matter), the *efficient cause* is the sculptor (the one who brings the statue into being), and the *final cause* is the idea that the sculptor wants to present in the statue, for instance, a historical person. We can say that the sculptor having a historical person on his mind acts toward this idea as to the goal of his sculpting activity. The idea then becomes the final cause. There is also a subtype of the efficient cause called the *instrumental cause*. In our example, the instrumental cause would be a chisel used by the sculptor. Even though the instrumental cause takes part in the creation of the effect, it has no power on its own to create the effect. Instead, it takes all its power from the prime cause. Furthermore, there may be many intermediate (i.e., secondary) efficient, instrumental or final causes in any chain of causes and effects.

Now, let's see if the compatibilists' solution to the problem of randomness in Darwinian evolution meets the demands of classical metaphysics. The idea they propose is evolution guided by God. This definitely introduces the final cause into nature. Exactly as in the classical non-evolutionary Christian approach, in theistic evolution nature has its goal, which is designed by God and leads all natural events, including the origin of species, to nature's ultimate end.

Yet, there are still two major problems that are not resolved in the compatibilists' view. First, on the classic approach, species are caused by the direct divine power, which is the only efficient cause capable of producing new forms (natures) in matter.[32] In contrast, the active principle in biological macroevolution (whether Darwinian or any other) is a combination of natural processes, such as generation combined with random variation and natural selection. This means that the efficient cause is conflated or even reduced "down" to the underlying material cause. Thus, the efficient cause disappears, because it is identified as the material cause.

The second problem is the lack of the formal cause. If every living being is to become something else (and this is the core idea of biological macroevolution), then a being never remains what it is. Thus, it never adheres to its own nature or substantial form. Instead, it always transcends its own form in order to become something else. In contrast, classical metaphysics teaches that the formal cause makes a thing what it is according to its nature. Moreover, no being wants to cease to be what

---

32. "God, though He is absolutely immaterial, can alone by His own power produce matter by creation: wherefore He alone can produce a form in matter, without the aid of any preceding material form" (*S. Th.* I,91,2, co).

it is; instead, it strives to preserve its own nature. Therefore, in theistic evolution, the formal cause is "swallowed up" by the final cause and thus disappears from the explanation.

Hence, in theistic evolution, there are only two out of four causes—the material and the final. Accordingly, in atheistic evolution, which lacks God as the end of all natural events, only the material cause is present.

We see here a bit of irony, because Christian philosophers who promote theistic evolution do so in opposition to the materialistic reductionism present in Darwinian evolution. Their goal is to reconcile the idea of evolution with the traditional philosophy of nature. Yet, they fall into the same type of reductionism that they eagerly combat among the materialists. The difference is that the materialists eliminate three causes and save only the material cause, whereas theistic evolutionists eliminate two causes and retain two others. Thus, theistic evolutionists differ from atheistic evolutionists only in the degree of reductionism adopted in their approach. In contrast, the proponents of classical metaphysics speak about four causes, which are preserved in biology only when transformation of species and universal common descent are eliminated.

※

I am well aware that many Thomists or philosophers of nature would not agree with the arguments against theistic evolution derived from classical metaphysics. The main reason for their rejection of these arguments is their abandonment of the realism of Thomistic philosophy and classical metaphysics. Moderate realism (the position between idealism and nominalism) was the cognitive position of Aristotle and Aquinas, and what they teach makes perfect sense only if we adopt the same epistemological attitude. The choice between realism and idealism is made by a philosopher before s/he begins any philosophical inquiry. Indeed, the type of his/her philosophy is largely determined by this antecedent choice. Some Thomists promoting theistic evolution may not even be aware of the position they have chosen; others, who are aware of the choice, may have accepted realism but do not really understand what it means, or are inconsistent, applying or ignoring it depending on whether or not they find realism convenient in a given context. Some accept realism but understand it as nothing but a form of idealism or nominalism that happens to be called realism. Yet, if we consistently accept realism, just as Aquinas did, we need to accept that species do not exist only in the mind; rather, they are real beings that exist (are realized) in each individual. If this is the case, then all of the arguments from classical metaphysics presented against biological macroevolution are true obstacles to accepting the evolutionary history of life. To put this concisely, Darwin's theory cannot be reconciled with classical metaphysics.

## e. Transformation of Species vs. Biological Macroevolution

On the margin of our metaphysical considerations, we should notice one problem surfacing in the discussions about the theory of biological macroevolution. The problem is not strictly logical or metaphysical, but methodological. Typically, when biological macroevolution is debated, the proponents try to show how one species (or organ, or protein, or gene) can be transformed into another and the opponents, how such transformation is biologically impossible, or how little evidence there is that it actually happened. Consequently, one of the core issues of the entire debate over evolution is species transformism—can one species (organ, gene, protein) be transformed into another? It is widely assumed that if the answer is "yes," then macroevolution is possible, while if the answer is "no," then macroevolution is ruled out. We need to notice, however, that the mere possibility of the transformation of species is not all that is needed for evolution to work.

Let's assume that evolutionists can actually present the evidence that one species has been transformed into another. What would it take to actually demonstrate such transition? The bare minimum would be to outline all biochemical transitory paths to create new proteins, new organs, systems, and body plans in such a way that all mutations and changes would not kill the animal on the way, but rather provide an overall survival advantage at each step. Of course, nothing even close to this has ever been shown.[33] We do not even know how to define or measure things like "survival advantage" or "evolutionary steps." However, we are assuming this hypothetical scenario for the sake of the argument. The question is, what does the transition from one species to another actually tell us?

---

33. Michael Behe, in each of his books, demonstrates that the attempts to present the origin of just one complex biological system are extremely rare in the scientific literature. But even these rare attempts do not really present any realistic biochemical scenario. Rather they produce some very general schemes of how such transition could be imagined. There is no such path for a random production of even a single functional protein.

The answer is that it shows us how to change one species into another once we have the idea and knowledge of both species. (The same applies not just to entire species, but to any complex, functional biochemical system, such as a bacterial flagellum or new kinds of proteins.) So, if we have "A" and "B," we can ask how to transform "A" into "B." We can ask this question, because we know what we want to obtain. We can try to measure the amount of difference between "A" and "B," look for transitory stages, check what new genes and proteins we need, which mutations could possibly generate those changes etc. All this inquiry is possible because we know what we are looking for. But in any realistic evolutionary scenario, we do not have two species. We have the species "A," but not the species "B", which does not exist in reality, nor in our mind. Therefore, we do not know where to go and what to look for. Evolution is a process that is not only supposed to change one species into another, but to create a new species altogether. And this is much harder; it requires not just foresight—the ability that no unintelligent process has—but powerful creativity which is more than just foresight.

This is why the real question for any macroevolutionary scenario is not how to change "A" into "B," but rather, what "A" should be changed into, and why? The evolutionary models of how evolution progressed are not satisfactory unless they can explain the creativity of the process, instead of merely showing its ability to transit from one form to another. When evolutionists present how, hypothetically, a new species could have emerged, they already adopt the final cause, which is the idea of the new species that they know from nature or fossils. Material processes, however, work within nature; they are unable to "go out" to search for novelties from outside of nature. Meanwhile, new species are packed with biological novelties. Evolution is unable to produce "B" not just because biochemical pathways from "A" to "B" are difficult or impossible, but because it does not even know what "B" is, or whether it is "B" or anything else.

The conclusion is that proving the possibility of the *transformation* of species is not the same as proving the possibility of the evolutionary *origin* of species. We can say that the transformation of species is a necessary but not a satisfactory condition of biological macroevolution. This means that if transformism is impossible then macroevolution is impossible as well, but even if transformism were scientifically and philosophically possible, it still would not mean that biological macroevolution can create new species. Even an exhaustive biological demonstration of how complex functional structure "A" could be transformed into complex functional structure "B" would not prove that this happened in nature, unless it was demonstrated that this could happen entirely by a blind evolutionary mechanism, without any foresight and knowledge of the structure to come within the available time.

## 4. Evolution and Science

Having presented the philosophical problems of biological macroevolution, we are moving on to a discussion of evolution in biology. There are many books on this topic, but few of those written by theologians or philosophers present the scientific response to Darwinian arguments with clear distinction from the religious and philosophical arguments. Books written by scientists, in turn, skip the theological part. Our goal is not to add anything new to the scientific debate on evolution, but rather to summarize just the main arguments in their most synthetic form. This review of science is a necessary appendix to the philosophical arguments constituting the main part of this chapter. By adding this part, we want to highlight the fact that correctly understood science will never contradict or diminish correctly understood philosophy and theology. Since biological macroevolution contradicts sound philosophy, it is no surprise that biological data stripped of theoretical (or ideological) interpretations also testify against biological macroevolution. Truth cannot contradict truth. Also, Pope Pius XII advised Catholic scholars that they should not ignore the science of their time:

> Catholic theologians and philosophers, whose grave duty it is to defend natural and supernatural truth and instill it in the hearts of men, cannot afford to ignore or neglect these more or less erroneous opinions. Rather they must come to understand these same theories well, both because diseases are not properly treated unless they are rightly diagnosed, and because sometimes even in these false theories a certain amount of truth is contained, and, finally because these theories provoke more subtle discussion and evaluation of philosophical and theological truths.[34]

Some of the compatibilists think that science and theology are completely separate realms, and therefore theologians should not enter biology and biologists should not make theological statements. This is how (as we noted above) some compatibilists effectively become isolationists.

---

34. Pope Pius XII, *Humani Generis*, no. 9.

# Evolution and Natural Knowledge

Their reasoning, however, does not take into account one important distinction: the realms of theology and science are separate when they speak about their proper objects (see our discussion of the three domains in I,1,a), but when it comes to the question of origins, both disciplines claim the answer, because the objects of the two disciplines become one. Theologians who know little or nothing about biology become an easy prey for the materialists cunningly using scientific language to promote materialistic philosophy.

## a. The Fossil Record

When Darwin proposed his theory, the general features of the fossil record were already known—namely, the sudden appearance of large groups of organisms and their unchanged persistence until extinction. Darwin was well aware at least of one problem: the lack of transitional forms.[35] Responding to this difficulty, Darwin claimed that the intermediate forms are fewer and do not last long, and thus their preservation is less likely. He hoped that the future would reveal fossils that would vindicate his theory of gradual change.

The logical problems of the Darwinian conceptual framework have already been discussed (in II,2). But even if we unreservedly accept Darwinian logic, there is still a serious paleontological obstacle against accepting his theory. Since Darwin's time, thousands of new fossils have been unearthed, but the leaps in the fossil record have not been evened out. On the contrary, the fossil record has turned out to be even more abrupt. In our times, Darwin's solution claiming the incompleteness of the fossil record raises the following question: How many fossils need to be discovered to overturn or confirm the discontinuity of the fossil record? If the number of samples was small, Darwin's response would be defensible. But fossils have been discovered in great numbers for the last two centuries. The number is enormous.

To better understand the problem, imagine that you are looking at an exit from a gigantic, multi-story parking lot. You may expect to see all kinds of vehicles leaving the parking lot, including such bizarre creatures as tractors, tanks, bulldozers, fire trucks, mobile cranes, coaches, etc.

---

35. Darwin asked, "Why, if species have descended from other species by insensibly fine gradations, do we not everywhere see innumerable transitional forms? Why is not all nature in confusion instead of the species being, as we see them, well defined?" (*The Origin of Species*, 171).

However, after ten minutes of observation, you have noticed just regular passenger cars and pickup trucks, and the same after another ten minutes, and another. Finally, after about an hour, you have spotted one small bus, and then again just cars and pickup trucks, for hours and hours. What are the chances that the lot is packed with those other types of vehicles? Similarly, the problem with the fossil record is that the linking forms should be the norm, and yet they are just rare and doubtful exceptions.

The notorious lack of transitional forms is one problem with the fossil record. Another is the fact that species represented by the fossils remain virtually unchanged throughout millions of years of their existence. This phenomenon is called *stasis*. If there are no transitional forms and the forms that exist do not change, we do not have a reason to assume transformation of species. At least, this is not where the fossils take us.

There is, however, another even greater challenge to Darwinian idea stemming from the *overall structure* of the fossil record. To present this issue clearly, first we need to see how Darwin interpreted fossil data.

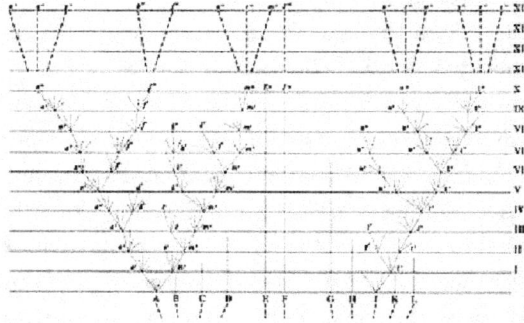

Diagram 12A. Darwin's original drawing of the "tree of life."

Diagram 12A shows the only diagram in Darwin's book *The Origin of Species*. Darwin presents a possible scheme of the emergence of species. It is worth noting that he does not speak about any real species. Instead of actual names there are symbols and letters. Though it is easy to imagine the Darwinian "tree of life," it is much harder to show the real evolutionary paths that would converge to a single ancestor in the distant past. This very fact makes the Darwinian concept very speculative and theoretical in nature. A critic would say that we can imagine many different things, virtually anything, owing to the overwhelming capacity of the human mind to collect and organize images and ideas. Yet, not everything that is imaginable can happen or did actually happen in natural history. The fact that the Darwinian imagery appeals to the human mind with great power does not, by itself, make it true or realistic. Since his imagery is coherent it might be true, but it may be delusive as well, and completely wrong in scientific terms.

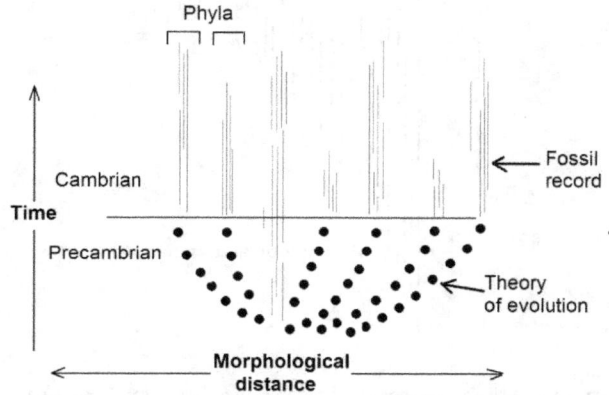

Diagram 12B. The actual structure of the fossil record.

Diagram 12B presents Darwin's idea with regard to the actual fossil record. We see separate phyla as disconnected from each other. Most of them emerged in the Cambrian explosion. The imaginative component of Darwin's theory is depicted by the black dots.

Since the higher levels of taxonomical classification give us a more general picture of life, we need to resort to them in order to see the *overall structure* of the fossil record. But, on the other hand, we should not go too high, because then we would miss some important features. An appropriate level to talk about the fossil record as a whole is the level of phylum. A phylum establishes a basic body plan for a given group. For example, the animals belonging to the phylum Arthropoda are invertebrate animals having an exoskeleton (external skeleton), a segmented body, and jointed appendages. Chordates are animals possessing a notochord, a hollow dorsal nerve cord, pharyngeal slits, an endostyle, and a post-anal tail for at least some period of their life cycles. Among Chordates are vertebrates, such as fish, amphibians, reptiles, birds, and mammals. Another example of a phylum is the Mollusca. These have a muscular foot, a mantle, and a round shell. All animals, plants, and fungi can be classified into about fifty phyla. The following diagrams (13A and 13B) depict the phyla as they appear in the fossil record over time.

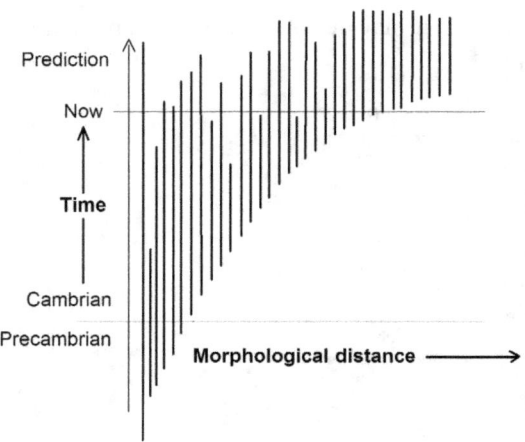

**Diagram 13A. The emergence of phyla as predicted by Darwin's theory.**

Diagram 13A shows how the fossil record should look if Darwin's theory were true. Since one phylum is supposed to give origin to another, there should not be simultaneous emergence of many different phyla, at least not in the early stages of the history of life. Instead, the first organism should constitute one species, one genus, one family, one order, one class, and one phylum. After a long period of diversification, some other species would emerge that would give rise to a new genus, family, order, class and finally a new phylum. The emergence of many new species would result in subsequent creation of new higher taxonomical groups. Later in the history of life, many phyla undergo the supposed diversification simultaneously, and thus they should diverge into more phyla in a shorter time than at the beginning. This is represented by the exponential acceleration of the emergence of new phyla. Yet, the actual record tells a different story (Diagram 13B).

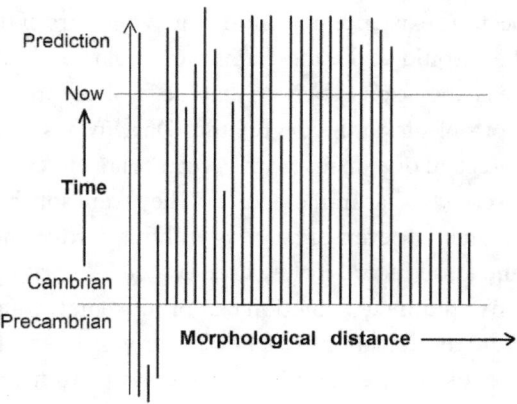

Diagram 13B. The emergence of phyla in the fossil record.

In the early history of life, we have just one phylum that includes the primitive single-celled organisms. Before the Cambrian explosion of life, there occurred another explosion (the Avalon explosion) that produced the so-called Ediacaran biota. These were multicellular organisms of mysterious anatomy and function. They elude any contemporary classification, and as such they are not good candidates for the supposed ancestors of the Cambrian phyla. The Ediacaran creatures completely disappear long before the Cambrian. Then, suddenly (in geological terms), the majority of animal phyla come into existence all at once. Many of them still exist today, but some became extinct. The Cambrian explosion (as well as a few other explosions on lower taxonomical levels that happened afterward) poses two major difficulties to the Darwinian tree of life.

The first difficulty is that phyla (as well as other taxonomical levels) appear suddenly, without any evolutionary history. The duration of the Cambrian explosion falls below the resolution of our time-measuring tools, which is roughly ten million years. This means that the event might have lasted ten million years or any shorter period of time within the ten-million-year span, including one year, one day, or one moment. In any case, even ten million years—the longest available time for the Cambrian explosion—is way too short for any realistic evolutionary scenario. To see how rapid this event was, we can compare the whole history of life on earth to a 24-hour day. The equivalent duration of the Cambrian explosion would be a maximum of four minutes.

If the fossil record were seriously incomplete (as Darwin claimed), we could suspect the existence of evolutionary ancestors of the Cambrian organisms. They would make the Cambrian explosion "less explosive." However, there is no reason to believe that the fossil record misrepresents the actual history of life to the degree that Darwin postulated. We find fossils of single-celled organisms from Precambrian epochs. We also find excellently preserved Ediacaran creatures. They were soft-bodied organisms, which means that even organisms without hard tissues should be preserved in the fossil record if they ever existed. According to the very logic of Darwin's argument, evolution can produce new things only very slowly, by minute modifications extended over vast intervals of time. In the Cambrian explosion, everything just pops up from nowhere. This clearly does not support Darwin's idea. Moreover, if the known fossil record properly represents the history of life and provides an actual picture of what happened, then it is *positive evidence* against the Darwinian interpretation of the history of life. The fossil record testifies against the Darwinian tree of life.

The second problem stemming from the overall structure of the fossil record consists of the fact that the tree of life suggested by Darwin appears to be upside down. Phyla, instead of branching out from one common trunk, appear diversified from the beginning and later seem to converge as a result of extinction. Thus, what comes first in biology are the different body plans, that is, completely new ideas of how to arrange the animal and plant parts in one functional organism. Later, within these broad categories, many variants show up, yet the general distinctive structures come first and remain unchanged. This means that life did not diversify from the bottom up (as Darwin imagined) but rather from the top down, as if it had been first prearranged and only later realized according to the previously conceived plan. This fact casts doubt on the overall logic of the Darwinian tree of life and moves us to a philosophical idea of design, rather than random variation and natural selection, as the source of new species.

## b. Homology

One of the most important Darwinian arguments for common descent is based on homology. The word homology comes from the Greek *homos*, meaning "same," and *logos*, in this context meaning "relation." Hence, the argument from homology appeals to the structural similarities between different organs or species. For example, let's consider the limbs of four different animals: the hand of a human, the leg of a dog, the wing of a bird, and the fin of a whale (see Figure 1).

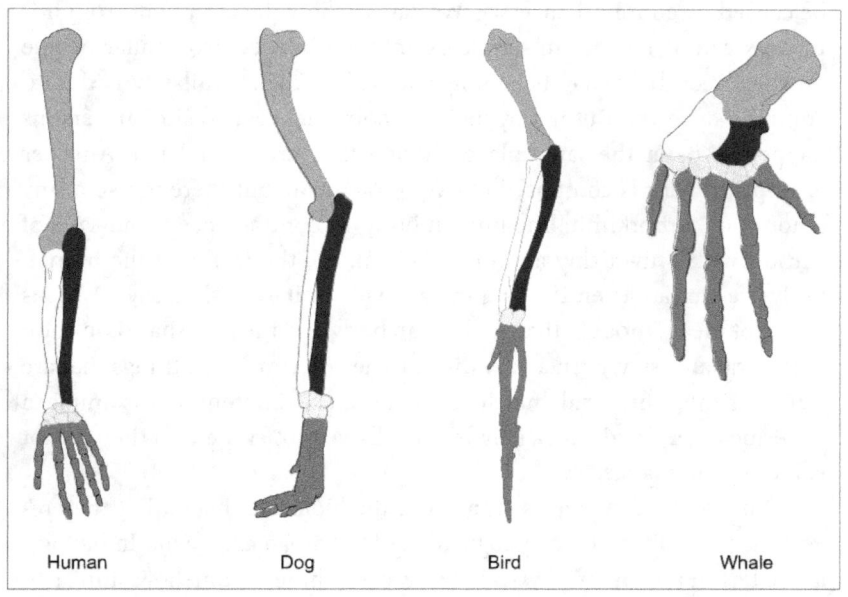

Figure 1. Skeletal structure of different limbs. (Source: Wikimedia Commons, https://upload.wikimedia.org/wikipedia/commons/f/fa/Homology_vertebrates-en-bw.png)

We immediately notice that the limbs have a similar structural pattern—humerus, radius, and ulna followed by a number of smaller bones. For Darwin this was clear evidence of the common ancestry of these animals. According to his theory, there must have been one organism long ago that developed the pattern that was then passed on to offspring, and over a long period of slight modifications it reached the current stage of diversity. Nevertheless, the original overall structure remained unchanged, most probably owing to its competitive advantage over other possible structures. Later, after we achieved our knowledge of genes, the same argument was transplanted into the field of molecular biology. Today, the most commonly cited argument for universal common ancestry is based on genetic similarities. For example, saying that a chimpanzee and a human exhibit about 99% genetic similarity can bring a rapid end to an everyday discussion of human origins—if the similarity is so great, those who doubt common ancestry cannot be right. However, there are a few fundamental problems with the argument from homology.

First, the fact that two things are similar *may* be but also *may not* be caused by common ancestry. We can see how the argument from homology can be illusory in simple examples: If we see two similar people on the street, they may be siblings, and then their similarity is due to common ancestry. But it may be just a coincidence: two similar persons happen to be at the same place, though they are not related. Another example: A fork is composed of over 99% metal, but there is also a tiny amount of carbon in it. The human body, in contrast, has about 18% of carbon with only a tiny amount of metal. So, the fork and the human body are similar, even if it is quite a small similarity—let's say 1%. This does not mean, though, that the human body and the fork share common biological ancestry, even a very distant one. There are also things that are built of similar material, but their form is quite different. For example, a watermelon, a cloud, and a jellyfish are all over 90% water, yet they do not share common ancestry.

Surely, these examples stray from the biological realm, but they reveal the problem in the very logic of the Darwinian argument. In biology, as well as in any other physical reality, everything is somehow similar to everything else. The fact that common ancestry is not the only possible explanation for similarities between different organic structures renders the argument from homology circular. This circularity can be seen in the biological literature, where evolutionary kinship is established by appealing to similarity, and similarity is explained by postulating evolutionary kinship.

This is even more evident if we enter the field of molecular biology. Academic journals and professional literature are filled with arguments "proving" the common ancestry of different animals and plants. However, what we find in these publications are almost always comparisons of gene sequences. The conclusion regarding common ancestry is reached on the assumption that greater genetic similarity corresponds to closer kinship. Very few authors attempt to show the real or even hypothetical evolutionary paths and thus actually prove how entirely new organs and biochemical systems could have been developed in the gradual process of random variation and natural selection. The few biologists who address these essential questions immediately move from real chemical compounds, reactions, and biochemical structures to the language of symbols and very general speculations. Although claims in favor of universal common ancestry are widespread among biologists, reliable evidence and realistic scenarios for major evolutionary transitions do not accompany the big postulates.[36]

### c. The Classical Explanation of Homology

Darwin was not the first biologist to notice similarities between organs or entire organisms. Similarities had been known since antiquity and throughout the Christian era, long before modern evolutionary theories came about. Darwin, however, provided a new interpretation of similarities, one which previously had not played any significant role. Before Darwin, similarities were attributed to common design rather than common ancestry. Organs are similar because they perform similar functions. Naturalists understood that the efficient performance of an organ requires a specific design.

To understand this better, we can compare biological designs to human designs. There are different types of buildings, but most of them require one thing—light. Designers (e.g., engineers or architects) have figured out that the best way to introduce light into a building is by means of windows. Hence, a school, a house, a church, an office building—all have windows. All windows share the same basic structure (typically a glass pane in a frame), but they also differ; one general design (solution) is adapted in different ways to different applications. This, however, does not mean that a school, a house, and a church have windows because engineers repeatedly modify one building into different buildings with different functions. Most buildings are constructed independently, with one purpose in the mind of the designer.

The same principle of "common design" rather than common ancestry is seen wherever human engineers are present. For example, flywheels, shafts, bushings, pistons, and cylinders are commonly used in all kinds of combustion engines. The same types of part are present in a motorcycle, a car, a truck, and a plane, not because one was derived from another but because engineers repeatedly use the same solutions in different applications. In fact, when something incorporates less novelty and more use of already known structures, its design is more reliable and

---

36. Michael Behe, in his groundbreaking work *Darwin's Black Box*, speaks about the abundance of gene sequence comparisons in scientific literature and a complete lack of realistic evolutionary scenarios of the emergence of complex cellular machinery. See *Darwin's Black Box*, chapter 8, 173–86.

efficient. Consequently, the ability of human designers to adapt the same basic solution to many different applications testifies to their contrivance.

It is similar in biology: homologies in the biological realm are explained by the activity of a mind that produces sufficiently optimal structures in order to maintain the existence of a given species. In Darwin's example of the limbs, the general structure of the different skeletons is the same because it provides both maneuverability and strength. And what would happen if, for instance, the bones were arranged in the opposite direction (i.e., many small bones attached through many joints to the body and then one robust bone at the end of the limb)? Surely, this kind of limb would have hardly any maneuverability and even less ability to grasp. The order of the bones in the limbs is not just the whim of evolution—it is an intelligent response to the physical constraints of our universe. The task of an engineer is to work out the best possible solutions within these constraints.

By the way, owing to the fact that the designer works efficiently when he adapts the same solutions to many different biological structures, any imagined shortcomings in biology are only relative, i.e., they may be demonstrated only in very specific conditions and during limited periods of time. When a healthy organism is considered and the whole context of its existence is taken into account, it is quite difficult to prove any real imperfection in biology.

### d. Survival of the Fittest as Tautology

British philosopher Herbert Spencer, after reading Darwin's *The Origin of Species*, coined the phrase "survival of the fittest." This refers to the Darwinian evolutionary principle that organisms which are better adapted to their environments will have an advantage in the struggle for life, and thus will survive and probably produce the greatest number of offspring. According to Darwin, if the same process occurs in subsequent generations, the organisms might permanently change toward greater adaptation. Over years, the idea of the survival of the fittest became the hallmark of Darwinian theory. The widespread use of the phrase, however, is rarely accompanied by a deeper reflection on its meaning. What does it mean that one organism is fitter than another organism? How can this be measured? Biologists speculate about different features that might provide survival advantage, but hardly any of their examples have ever been tested.

For instance, a biologist may think that anaerobic bacteria are more likely to survive than aerobic bacteria in the absence of oxygen. This is a proposition, but not a fact. Since virtually any trait in any organism may turn out either advantageous or disadvantageous in different conditions, biologists do not have any better criterion for establishing the "fitness" of an organism than the number of offspring that the organism produces. They assume that the fittest are those who will propagate successfully and leave the largest number of offspring. This is why *the fittest* are commonly defined as *those that produce more offspring*. As the fittest survive in subsequent generations, they should establish a greater population that will last longer. When biologists encounter a "successful" population (and, in this context, any living population must be considered successful), they can assume that it consists of the fittest, because only the fittest would create a successful (i.e., living) population. Hence, the members of a "successful" population are considered the "fittest" because they survived and created a larger population, yet, they survived and created a larger population, because they were the fittest. The popular saying therefore ends up in a tautology: those that survive are the fittest and the fittest are those that survive. After all, it is obvious that the fittest would survive. This, however, does not prove that the evolutionary process would continue and produce any new "fittest" organisms. The survival of the fittest does not explain their "arrival." Instead, it assumes their existence.

We also should notice that producing a greater number of offspring is not necessarily advantageous for the population. In fact, since individuals within the same population tend to use the same resources, more offspring leads to greater competition among members of the same population. The survival of the fittest leads to the extermination of the fittest, such that the whole process becomes self-limiting. Survival of the fittest is not a good candidate for producing the one-directional changes demanded by macroevolution.[37]

### e. Epigenetics

In the neo-Darwinian account of evolution, mutations in DNA sequences produce new genetic information that is needed to build entirely new

---

37. Michael Behe presents biological evidence for this general claim. It turns out that mutations making organisms "better fitted" to one niche simultaneously make them less flexible and therefore more susceptible to extinction when they are forced out to other biological conditions. See *Darwin Devolves*, 149–52, 161–66, 177–83.

organisms. The neo-Darwinian approach is based on the assumption that DNA is a sufficient source of information for building and governing the biological processes of a living being. Consequently, if we alter the genes, the whole organism should change and eventually produce a completely new form. There are a few reasons why the neo-Darwinian mechanism boils down to genetic mutations and natural selection. First, genes constitute a reservoir of information, and the standard process of gene translation and expression in organisms is relatively well understood and well-defined; thus, this is an attractive place in which to look for the cause of evolution. Second, alterations of genomes indeed produce some changes in organisms. Finally, those changes can sometimes be inherited and passed onto offspring. For these reasons the neo-Darwinian mechanism seems to be the best candidate for explaining the increase of new forms in the history of life.

Today, however, the standard neo-Darwinian account of evolution encounters a new problem, originating from the field known as epigenetics.[38] A number of experiments performed during the last few decades have demonstrated that there are sources of information in organisms other than genes (hence the name *epigenetics*, from Greek *epi* meaning "upon, on top of, in addition to"). Since the 1940s, it has been found that some embryos undergo a few developmental phases after their nucleus has been removed, that is, without their nuclear DNA, which is the main source of genetic information. Another example is the specialization of cells. Despite the fact that each cell of one organism contains the same genetic information, they specialize according to different functions required by different types of tissues and organs. This demonstrates that there must be something else, some other information determining how cells specialize into different types. We know that genetic information plays a crucial role in regulating the synthesis of proteins. Yet, DNA does not contain information about the spatial arrangement of those proteins. In fact, every organism has different levels of organization, and each of them requires new information. As Stephen C. Meyer puts it, "Other sources of information must help arrange individual proteins into systems of proteins, systems of proteins into distinctive cell types, cell

---

38. Sometimes the term epigenetics refers to the study of the embryonic development of organisms and specifically the developmental regulatory genetic mechanisms. Here we use it in a more current sense, as the term concerning the study of cellular information stored outside of the chromosome.

types into tissues, and different tissues into organs. And different organs and tissues must be arranged to form body plans."[39]

Epigenetic information plays a crucial role in the development of an organism. To see how this works, we can compare a growing organism to a construction site. Proteins are like basic building blocks—bricks, wooden beams, steel rods, etc. Genes tell the cells what types of basic materials are needed as well as when to deliver them and how much to deliver. But they do not say how to arrange them into three-dimensional structures. Developmental biologists have identified several sources of epigenetic information stored, for example, in cytoskeletons in eukaryotic cells and in membrane patterns. This information exists, for instance, in a form of spatial arrangement of cellular parts that is present in the organism from its inception. As far as we can tell, most epigenetic information is neither produced nor altered by the genetic information.[40] If this is the case, the gene-centered approach to life is essentially outdated and inadequate. Specifically, it turns out that the neo-Darwinian mechanism cannot account for all changes that would be needed to produce new forms of life. It can change the amount of and sometimes slightly modify the type of basic building blocks (i.e., proteins), but it neither creates entirely new building blocks nor tells the cell how to arrange them in space. This is why no matter how many mutations (random or guided) occur in DNA, they will not create a new species. The neo-Darwinian mechanism is simply unable to account for the changes that are needed to create a new form of life. It follows that biology currently does not have any credible mechanism for explaining supposed macroevolutionary changes. For the same reason, the similarities and dissimilarities between the genomes of different species do not say much about their supposed evolutionary relations. Indeed, even if humans and chimpanzees had 100% identical genomes, they could still be entirely different species because of the epigenetic differences.

## 5. Intelligent Design and Neo-Darwinism

Thus far we have presented arguments related to the scientific discussion about biological macroevolution. In recent decades a number of scientists who disagree with the far-reaching claims of the neo-Darwinists have

---

39. Meyer, *Darwin's Doubt*, 276–77.
40. Meyer, *Darwin's Doubt*, 282–84.

come up with an alternative theory of the origin of species. It is called the theory of intelligent design (ID). It has two aspects: a negative one that presents the difficulties of the standard (neo-)Darwinian account of evolution, and a positive one that presents the positive case for intelligent design in biology.

The edge of ID criticism of evolution is directed against the neo-Darwinian mechanism of random variation and natural selection. Scientists supporting ID demonstrate the inability of the neo-Darwinian mechanism to produce new genetic information, new proteins, or complex biochemical systems. It is to be noted that ID, by itself, does not exclude universal common ancestry or the transformation of species. What it does exclude is the evolutionary claim that merely the interplay of chance and necessity (the two basic elements of Darwinism) explain the origin of *all* novelty in biology. ID recognizes the limits of the neo-Darwinian mechanism and postulates a third factor—intelligence, that is, a power endowed with foresight, capable of acting according to a preconceived goal. According to ID proponents, only this kind of cause can account for the multiple information-rich, irreducibly complex structures found in living beings.

There is a great deal of a debate over ID and its status in the science of biology. Some of the issues will be addressed in later subsections of this section (II,5,b–i). At this point, however, we should ask a more general question: Is it at all possible that one fundamental theory can be replaced by another? Since neo-Darwinism is not just one particular theory but more like a paradigm of doing biology, is it even possible to replace it with another paradigm?

Thomas S. Kuhn proposed that development in science does not happen due to the accumulation of new data that slowly modifies the established theories. On his view, scientific progress is a result of revolutions that happen after periods of "normal science"—during which periods "anomalies" do bother scientists but are not enough to overturn scientific paradigms.[41] Today we see that new discoveries in biology are harder and harder to reconcile with the neo-Darwinian mechanism of evolution.[42] This may indicate that the scientific community is entering

---

41. Thomas S. Kuhn, *The Structure of Scientific Revolutions*.

42. In 2016 the Royal Society in London organized a conference, entitled "New Trends in Evolutionary Biology: Biological, Philosophical and Social Science Perspectives." At the conference, world experts in different fields of evolutionary biology shared their doubts and questions regarding the neo-Darwinian synthesis. The very

what Kuhn calls the third phase, that is, the period when the paradigm proves chronically unable to account for anomalies.

If the shift from neo-Darwinism to intelligent design in biology were to happen, it would be analogous to the transition from classical mechanics to general relativity in physics that took place a century ago. When Einstein came up with his revolutionary theory, he did not entirely disprove classical Newtonian mechanics. He showed that the explanatory powers of Newtonian mechanics are limited—Newton's theory works well with smaller distances and lower velocities. To explain phenomena occurring on a galactic or intergalactic scale, we need a broader theory. Similarly, ID does not say that random genetic mutations and natural selection do not play any role; rather, their roles have been vastly exaggerated in evolutionary biology. They can explain small changes such as an increase in average beak size in a population of finches or bacterial resistance to antibiotics.[43] To explain larger phenomena, such as the origin of entirely new organs, functional genes, or biochemical systems, we need a broader theory. ID claims to be it. In what follows, we will present the basic scientific arguments for intelligent design and then move on to philosophical questions regarding ID, such as whether it is science and whether or not it is based on the "god of the gaps" argument.

### a. "Probabilistic Miracles"

Before we present the basic arguments for intelligent design, we need to clarify what is meant by chance, necessity, and information within the context of the neo-Darwinism vs. intelligent design debate. In order to

---

fact that various alternative mechanisms were proposed confirms that evolutionary biologists see the limitations of the "classic" neo-Darwinian explanations. See the article by the Evolution News team, "Why the Royal Society Meeting Mattered, in a Nutshell," for more information.

43. The variation of beaks in the population of the Galapagos Islands, which became a flagship of Darwinian evolution, today, after more advanced studies, appears to prove the opposite: the limitations of the Darwinian mechanism rather than its creative capacity. Michael Behe quotes many examples of cases where evolutionary changes, including the sizes and shapes of beaks in finches, are produced by destructive genetic mutations (devolution) rather than creation of new genetic information (*Darwin Devolves*, 150–52). Also, another study argues that epigenetic information may be responsible for the changes in finches' beaks. See McNew, "Epigenetic Variation between Urban and Rural Populations of Darwin's Finches."

do it we will refer to a few examples of how these three factors can be recognized in daily life as well as in the science of biology.

For decades, some of the critics of Darwinian evolution maintained that the probabilities of even the simplest evolutionary events are too low by far to be attributed to pure chance. For example, Fred Hoyle (1915–2001) calculated that the probability of the random occurrence of enzymes necessary for the production of a single amoeba is 1 in $10^{40,000}$. (By comparison, the probability of randomly drawing a specified atom in the visible universe is 1 in $10^{80}$.)[44] Such extremely low probability is enough to make any reasonable person doubt that the event would occur by pure chance. Yet, there is a problem with the logic of this kind of argument against evolution.

To understand why, let's imagine a box with 26 cards, each marked with one natural number between 1 and 26. Now we randomly draw 10 cards from the box (after each draw we return the card so the total number of cards is not reduced). We do this three times and obtain the three following sequences of numbers:

a. 1,4,20,9,17,4,10,5,11,12

b. 7,7,7,7,7,7,7,7,7,7

c. 7,15,20,15,19,3,8,15,15,12

Any person who is aware of what is going on would immediately question the randomness of draw (b). Obtaining "7" ten times in a row is so unlikely that to suggest that it happened by pure chance is simply unacceptable. The doubting person would do the math and conclude that the probability, which equals $1/26^{10}$ (or roughly 1 in $10^{14}$), is just too small to allow the event to occur by pure chance. Yet, according to the same math, the probability of each of the three events is exactly the same; event (b) is not less probable than event (a) or event (c). Therefore, the argument from "low probability" does not really help to rule out the randomness of event (b) while preserving the possible randomness of (a) and (c). The same problem applies to biology: the formidably low mathematical probability of virtually any significant evolutionary event is not a decisive argument for ruling out an occurrence by pure chance. In this context, mathematical calculations simply are not the proper tool. The reason is that they are abstract; they account for theoretical objects,

---

44. Hoyle and Wickramasinghe, *Evolution from Space*, 24, 130. Hoyle's calculations were later confirmed by more advanced research; see Meyer, *Signature in the Cell*, 213.

not real events. When we transition from reality to mathematics, some essential features of the universe may be lost. And this is what happens when we estimate probabilities. The meaning of the numeric symbols is lost. Our estimation accounts only for the occurrence of an abstract object—a "number"; it does not include the important message about which number it is. Hence, we need to find some other way to distinguish between random and nonrandom events.

The thorough answer for how to do this has been recently provided by William Dembski, one of the founders of intelligent design theory. Before we present his "explanatory filter," however, we should take note of what a Polish Jesuit, Piotr Lenartowicz, proposed in his 1993 paper "On Probabilistic Miracles."[45] Lenartowicz speaks about the naive belief of some scholars that whatever is allowed by probabilistic theory (i.e., an event with probability greater than zero) may actually happen in the physical world.

One example is a locomotive that stops owing to dissipation of energy (running out of fuel, air friction, the resistance of the track, etc.); immediately after stopping, thanks to an accidental convergence of particle motions, the locomotive starts moving backward. The probability of this kind of event is greater than zero, and thus it cannot be ruled out mathematically. Nevertheless, claiming that anything like this could actually happen is like believing in a probabilistic miracle. Another example is the famous "infinite monkeys" scenario, in which a monkey, out of countless monkeys randomly hitting keys on keyboards, comes up with one of Shakespeare's poems. The proponents of probabilistic miracles maintain that anything can happen as a result of chance. Yet, our experience infallibly proves that this is not the case. These kinds of events do not happen in the real world. Therefore, we have grounds to suspect that some events, even if their statistical probability is greater than zero, are impossible. For many people, this conclusion sounds reasonable, but to explain why is not so easy.

In the "infinite monkeys" scenario, which is one type of probabilistic miracle, the postulate is that random treatment of a keyboard can yield an intelligible text. However, if the choice of characters is truly random, each character should appear roughly the same number of times. Assuming that our keyboard has just 10 numeric keys and 26 letter keys,

45. Lenartowicz, "O 'cudach' probabilistycznych, czyli fakt selekcji i odmowa poznania tego faktu" ["On probabilistic 'miracles,' that is, the fact of selection and refusal of recognition of this fact"].

the probability of randomly hitting any key is 1/36. If the keyboard gets hit less than 36 times, the probability of getting all the characters in the sample is zero. But after the keyboard is hit a thousand times, we would expect that each letter and number will occur more than one or two times. The number of hits for each key tends to be 1/36 of the total number of hits. For instance, if there were 36,000 hits, then each key should get hit about a thousand times. We would get a string of 36,000 characters, which is the length of an average poem. But the problem is that no intelligible text corresponds to an equal distribution of characters.

In fact, it is the opposite—every human language follows rules that disrupt the equal occurrence of characters. In our example, the numeric keys constitute roughly one third (10/36) of the total number of keys. Yet, in the vast majority of human writings, numbers appear much less frequently than once for every three characters. Similarly, vowels constitute less than 20% of the alphabet (5/26), but in a written English text they account for about 40% of all letters.[46] Also, there are letters such as $x$ that appear less often than other letters. If the way in which the keys were hit were truly random, all the patterns that characterize English actual text would be gradually eliminated by randomness. Believers in probabilistic miracles claim that if we try long enough, finally we will obtain any result. But the truth is the opposite: the more hits there are, the more similar is the number of hits for each character, and the more the result strays from any meaningful outcome.

On the other hand, if the result of hitting the keyboard revealed a great divergence from the expected outcome (such as an increase of vowels from the expected 20% to an actual 40%), a statistical test for randomness would show that the event was not truly random. Hence, either hitting the keyboard is random, and then it does not produce an intelligible text; or it produces an intelligible text, but then it is not random.

The more general conclusion from this example is that random events do not obey rules, whereas all languages (like all sources of specified information) follow rules and therefore linguistically meaningful statements cannot be created by chance. Lenartowicz's paper gives us a hint as to why probabilistic miracles cannot happen. Still, there is a greater question that remains unanswered, namely, how to distinguish events (or structures) that are intelligently designed from random and necessary ones. William Dembski's explanatory filter comes to our aid.

---

46. See "Consonants" in *Encyclopedia Americana*, vol. 3, 449–53, 452.

## b. The Explanatory Filter and Specified Complexity

Based on these considerations we can now define three types of events: (1) random (i.e., undetermined, unplanned),[47] (2) necessary (determined), and (3) designed (planned).

Random events are the hardest to grasp, because in nature everything seems to be determined by physical laws. If I throw fair dice in a fair way, the outcome is considered random. However, if we were able to determine the initial conditions and calculate all involved forces (friction, gravity, impetus, etc.), we could predict the outcome of a throw because it is determined by these factors. We call a throw of the dice or any other lottery of this kind a random event because we are unable to predict the outcome.

A better idea of randomness comes from the subatomic or quantum level. For example, the position of an electron within an atom is considered random because it is unpredictable for us. But at the same time, whenever this kind of randomness takes place in nature, the prediction of the particular event is quite irrelevant for us because what we need (in science or practical life) is just an overall picture or statistical probability of the event. For instance, in the case of the electron, it does not matter where it happens to be at a particular moment, or in which direction it is heading. What matters is that it is in motion, and owing to its motion it does not collapse into the nucleus, and by not collapsing into the

---

47. We are aware of the distinction proposed by Stephen Barr in his 2005 polemic against Cardinal Christoph Schönborn. Barr challenged the cardinal's 2005 article in *The New York Times* because, according to Barr, Schönborn confused two meanings of randomness. In physics (as well as in biology, as Barr implies), random means simply "uncorrelated," which is not the same as purposeless or—by extension—meaningless, senseless, etc. According to Barr, the use of "randomness" in science does not have the major philosophical baggage that it usually has in philosophical texts advocating the absolute dysteleology and accidental nature of physical events in the universe. Based on this distinction (uncorrelated vs. purposeless), Barr defends the use of the notion of randomness in evolutionary biology, even if he does not agree with far-fetched conclusions about an overall lack of teleology in nature. However, Barr's distinction seems to confuse the very idea of randomness on the one hand, and the use or implications of this idea on the other. According to Aquinas (for instance), a random event happens when two independent chains of causes and effects cross. Thus, for Aquinas, lack of correlation seems to constitute the essential aspect of randomness in general. It seems that in any use of words such as "random" and "chance," the core idea is some kind of independence of the events from each other, regardless of whether the words are used in a scientific or philosophical setting. And it is to this core idea of randomness that we refer in our considerations. See Schönborn, "Finding Design in Nature," and Barr, "The Design of Evolution."

nucleus it preserves important qualities of matter.[48] Yet even though the subatomic level gives us a better idea of randomness, it does not follow that subatomic events are truly random. We do not really know whether the movement of an electron (or any quantum phenomenon) is truly random or just inaccessible to our scientific method. As in the case of a lottery, we consider them random because we are unable to predict the positions of particles. The difference is that in lotteries we encounter only practical constraints, whereas quantum events are unaccountable for us in principle (Heisenberg's uncertainty principle). Hence, we should conclude that we do not really know whether or not there are any truly random events in nature. It is possible that all events in nature are actually determined, and we simply call "random" those events whose outcome we are unable to predict.

In contrast, events regulated by physical laws are called necessary. Since these kinds of events can in principle be investigated and explained by means of the scientific method, they constitute the proper object of science. Laws are detectable thanks to the regularities they create in nature.

The third type of event is the designed event. In today's debate, all major parties agree that in biology we deal with information and that many biological structures look as if they were designed. The points of controversy are two.

First, where did biological information come from in the first place? The proponents of neo-Darwinism say that information (such as genetic information) may be generated by the combination of random and necessary events (random genetic mutations and natural selection). The proponents of intelligent design say that information can be passed on and communicated by necessary processes (such as the process of generation in biology), but must originally come from a mind. In other words, there is no source of specified functional information other than intelligence.

The second controversial issue is whether many biological structures only *look as if* they were designed or were *truly* designed. This boils down to the question of whether chance and necessity can imitate the outcome that we normally attribute to intelligence. Again, the proponents of neo-Darwinism say, "Yes." In contrast, the proponents of intelligent design say that if an event or a structure bears the characteristics of being designed,

---

48. This is true under Bohr's model of atom. The Bohr model has been considered outdated for a long time, but for our argument it is irrelevant what exactly happens with the electron. The point here is that the subatomic particles behave in particular ways, and on a large scale they produce order that we see in nature.

it must have been actually designed, because we do not know of any cause that produces designs other than an intellect.

The explanatory filter does not settle these questions, but it enables us to establish which events or structures bear characteristics of being designed. Consequently, if one agrees that when something looks designed then it must be designed, one will agree that the explanatory filter reveals which events or structures are actually designed. If one does not agree, then the explanatory filter will only allow one to distinguish events that look designed from those that do not look as if they were designed. These distinctions are important in enabling us to understand where the explanatory filter takes us and where it does not.

The filter consists of three stages. In the first stage, we need to establish whether a given event is random or there is a law that governs it. Laws thrive on replicability, yielding the same result whenever the same antecedent conditions are fulfilled.[49] If something can be explained by a law, then it should not be attributed to design. Events attributed to laws do not go to any further stage, as they are already explained.

In the second stage, we are left only with events that are either random or designed. We look at the probability of the event and determine if the event can be reasonably attributed to chance. If the probability is high enough to allow for the event occurring by chance, then we should attribute the event to chance. Otherwise, we should take it to the third stage. (On what can be considered a "too low" probability, refer to II,5,c.)

In the third stage, we are left only with events that are extremely improbable. If an extremely improbable event is specified, i.e., matches an independent intelligible pattern, we can conclude that it is designed. Diagram 14 depicts the three-stage action of the explanatory filter.

---

49. Dembski, "The Explanatory Filter."

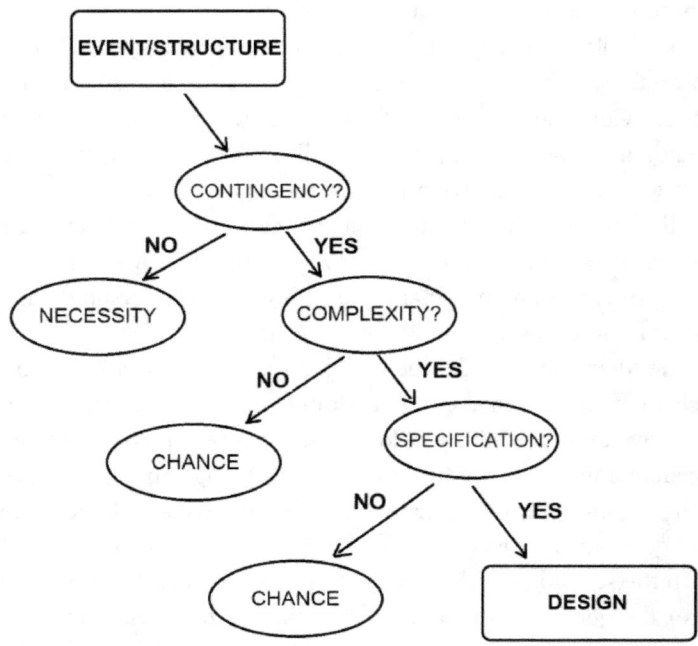

Diagram 14. Dembski's explanatory filter.

To explain how the explanatory filter works in practice, we will use two examples—one from our daily experience, another from biology.

When we see a statue or a monument representing a national hero or a saint, we never think that the stone or metal was accidentally arranged into a statue; rather, we conclude that it was designed and purposefully sculpted. This is a spontaneous judgment made by every conscious person when seeing a monument. Thanks to the explanatory filter, our judgment can be turned into a rigorous argument.

Let's assume that the monument is the bronze Lenin in Seattle's Freemont. In the first stage we ask if there is any necessity that makes bronze assume the shape of Lenin. Surely, no physical law makes bronze adopt this particular shape. We do not find there any regularity or repetitive pattern that could account for this particular shape. Neither can we describe it by means of a simple mathematical formula. In contrast, the shape of the earth is determined by physical laws. Any material object possessing sufficient mass will assume the shape of a sphere. And if an object rotates fast enough, it will be flattened by centrifugal force. And this is precisely what we observe in the shape of the earth. Two laws—gravitation and inertia—account for its basic shape. Hence, the shape of the earth stops in the first stage with the conclusion that it does not need to be designed. Lenin's statue, however, passes to the second stage, when we ask, how likely is it that the block of bronze would assume this particular shape? Since there is a virtually infinite number of possible arrangements of bronze particles, this particular arrangement is highly improbable. The statue passes on to the third stage, when we check if there is an independent pattern (a blueprint or an idea in our minds) that specifies the block of bronze with this particular shape. In this case, the independent pattern is a historical person, namely Lenin. We can learn about his appearance, character, and ideas from sources other than the monument itself. This is how we build the independent idea of a historical person in our minds. This idea matches the highly improbable and undetermined physical structure, and thus the statue exhibits specification. Events or structures that are highly improbable and specified are designed, and therefore Lenin's monument in Seattle is designed. Why is it designed in this particular way? Who designed it? Is it the best possible design? All these questions are irrelevant to our conclusion.

A biological example of detecting design comes from genetics. Let's assume a short gene **A** of a length, for instance, a thousand base pairs. Let's assume that it encodes a protein **B** that performs function **C**. The function of the protein is determined by the shape of the protein, which, in turn, is determined by the sequence of base pairs in the gene. In the first stage of the explanatory filter, we ask, is the sequence of the base pairs in a gene determined by any law, or can they assume any other sequence? Indeed, we do not find any law that would determine a particular sequence of the base pairs because the phosphate-deoxyribose backbone (i.e., two strains constituting the "skeleton" of the DNA double helix) can accept any of the four nucleobases (A, G, C, T) at any attachment location. In other words, the sequence of nucleotides in the DNA double helix is not determined by any chemical, physical, or structural necessity. Neither do we find any pattern, repetitive scheme, or mathematical formula that would describe the actual sequence of the base pairs (i.e., combinations of two nucleobases) in a gene. Hence, we conclude that there is no law accounting for the particular form of our hypothetical gene **A**.

We move on to the second stage, where we consider the probability of obtaining gene **A**. Since each base pair may have four variants (A–T, T–A, C–G, G–C), the probability of **A** emerging by chance is $1/4^{1000}$, or roughly $1/10^{602}$. This is approximately equivalent to randomly picking up a single marked atom out of the entire universe seven times in a row. Obviously, this kind of event qualifies as extremely improbable. Hence, we move to the third stage, in which we consider the existence of an independent pattern that indicates specification in the structure of the gene. In fact, we find specification, because it is precisely owing to the order of nucleobases in gene **A** that it encodes protein **B** with function **C**. Both **B** and **C** specify **A**. We see, therefore, that the gene is not an extremely improbable random structure but rather a designed structure.

No wonder that contemporary molecular biology, even if completely infiltrated by the neo-Darwinian paradigm, cannot avoid the language of mathematics, information theory, and computer science. An analogy that naturally comes to our mind compares the genomes to a computer program. In both cases we deal with information that is translated into specified useful functions. And as much as we do not claim that a computer program is an effect of randomly tossing out "zeros and ones," we also should not accept the idea that the genomes emerged through random mutations and natural selection.

### c. A Universal Probability Bound

When moving from the second to the third stage of the explanatory filter, we used expressions such as "extremely improbable" or a "much too low" probability. These expose the filter to a charge of employing imprecise terms that are generally not welcomed in science.

To see what kind of probability is too low to be reasonably attributed to chance, we need to refer to a concept called the universal probability bound. Today, thanks to knowledge from different sciences, we can estimate the lowest probability of a possible event in the observable universe. First, we know the time of the existence of the universe. This is slightly less than $10^{25}$ seconds. Second, we know that the observable universe contains about $10^{80}$ elementary particles, and third, we know that physical events (transitions from one physical state to another) cannot occur faster than $10^{45}$ times per second. Hence, the number of all physical events that could have happened throughout the entire history of the universe is about $10^{150}$. This is the total of probabilistic resources we have at hand. Consequently, any event whose probability is lower than 1 in $10^{150}$ should not be expected to happen in our universe throughout its entire existence. Since the probability of an accidental formation of a very short gene is approximately $1/10^{602}$, we are justified in saying that the probability is "much too low" to attribute the event to chance. In fact, even the relatively simple evolutionary events proposed in neo-Darwinian theory dramatically exceed the universal probability bound.

For a few decades now, different scholars have been calculating probabilities of these events. Some of them do not immediately exceed the universal probability bound, for example reaching half of this value or one-third. But these are still very low-probability events, because they would need to take place many times (usually millions if not billions) in order to account for the entire history of life.[50]

### d. Where Does the Explanatory Filter Take Us?

To conclude our discussion of chance, necessity, and design, we should make three final remarks regarding the power of the explanatory filter. First, we said above that the filter does not tell us whether an event or structure is actually designed or only *looks as if* it is designed. However, there is a major problem with allowing that things that look designed may actually be not designed. All science is based on the assumption that nature does not deceive us. If we accept the idea that something looks different from what it really is without providing strong evidence, we undermine the core principle of all natural science. The only way for us to study nature is through its appearances, so if the universe was deceptive, we could not study it.

Most people agree that some things in nature indeed look different from what they are. However, we can say this only if we have an independent way of establishing what the thing *is* and comparing it to how it *appears*. For instance, an optical illusion called a mirage makes us think that there is water on the surface of hot asphalt far ahead of us. We know

---

50. For example, one study shows that the probability of random occurrence of protein sequences adopting functional enzyme folds is not higher than 1 in $10^{77}$. This is roughly a probability of picking randomly a marked atom from the entire universe. But we are talking here about just one functional protein. Any organism requires many of them. See Axe, "Estimating the Prevalence of Protein Sequences Adopting Functional Enzyme Folds."

that this is an illusion because we detect no water while we move forward. If we do not have a way to prove an illusion, we need to believe that things are what they appear to be. In the case of genomes, neo-Darwinists *assume* that genes emerged through random genetic mutations and natural selection, but they cannot prove it. At the same time, they cannot disprove that they were designed. Therefore, there is no evidence that genes only look as if they were designed, and consequently we should assume that they are actually designed.

Second, the explanatory filter detects events or structures that are designed. However, intelligence can imitate effects that could be attributed to other causes. For example, criminals may try to conceal the traces of their crimes and thereby make them look like accidents. Because of this ability of an intelligent agent to imitate chance or necessity, not every intelligently designed event is detectable. For this reason, even if an event or structure looks random or necessary, it could be intentionally designed. The explanatory filter will not tell us whether an event or structure is designed unless it bears the characteristics of being designed. To see this clearly, let's refer again to the three numeric sequences from our discussion of probabilistic miracles.

a. 1,4,20,9,17,4,10,5,11,12

b. 7,7,7,7,7,7,7,7,7,7

c. 7,15,20,15,19,3,8,15,15,12

Sequence (a) represents a random event. Since sequence (b) can be described by a simple mathematical formula, it represents a pattern that amounts to a necessary event. Finally, sequence (c) represents design, because if we assign the numbers 1 to 26 to each letter of the alphabet, the sequence contains a piece of information: "go to school." Sequence (c) encodes a message and thus is specified. This is why, even though the mathematical probability of each sequence is the same, the first can happen by chance, the second can be produced by a law, and only the third is and must be designed. (Of course, since chance and law-like outcomes can be imitated by a designing mind, here also sequences (a) and (b) are designed by the author to represent different types of events.)

Finally, if an event or structure contains specified information, this fact can hardly be hidden. This is precisely what encrypting machines, such as the Enigma, try to do. They attempt to conceal the independent specification and incorporate the information into patterns that look random or necessary. But specified information can never be reduced to these patterns (without destroying the information). For this reason, if specified information is present in nature, it must be detectable.

### e. An Agent, a Law, and Chance

On the margin of our discussion of different types of events we should point out one commonly overlooked problem of evolutionary thinking. Oftentimes, both in popular discussions and academic presentations, we hear that evolution did this or that, as if evolution was some kind of agent. In such cases, evolution is personalized and made to act almost like an ancient goddess or a mysterious force with consciousness. This tendency of personalizing evolution is unavoidable precisely because evolutionary science attributes to evolution the effects that are normally produced by agents endowed with reason and the ability to work towards a predetermined end.[51]

Since evolution works by the neo-Darwinian mechanism of random variation and natural selection, it is a law-like process characterized by necessity. Laws create unsophisticated, repeatable and predictable effects. For instance, gravity draws things endowed with mass to each other. Thus, we say that gravity keeps us on the surface of the earth, or that gravity keeps the earth in orbit. When it comes to chance events, their outcomes are unpredictable, but at the same time they are meaningless. Chance events are irrelevant for nature; they do not produce anything nor explain any phenomena. If chance events are regarded in a large scale then they may produce patterns, but in such cases chance is transformed into a law. For example, random movements of particles in a gas are responsible for its energy, but it is not the random movements (even if they are actually random) of particles, but the average outcome produced by billions of particles, that taken as a whole informs us about the energy that the gas has. Truly random events in nature either do not exist or are

---

51. Pope Benedict XVI indicated this problem in one of his homilies: "Not only popular writing about science, but also scholarly scientific texts about evolution often say that nature or evolution did this or that. Here the question arises: Who in fact is Nature or Evolution as an acting subject? There is no such person! If someone says that Nature does this or that, this can only be an attempt to summarize a series of processes in a subject that, however, does not exist as such. It seems obvious to me that this (perhaps indispensable) linguistic expedient contains within it a momentous question." See Horn and Wiedenhofer, eds., *Creation and Evolution*, 162–63.

rare and exceptional and thus they do not serve as an explanation for things happening in nature. Laws explain most natural events, but they can only account for repetitive and predictable patterns. Evolution, as it is presented in science, is a kind of law and therefore it cannot explain the origin of entirely new things or events that do not act according to simple predictable patterns. The emergence of entirely new organs and species transcends the order of natural laws. And this is why, at the end of the day, either evolution is presented as an agent, a personal force that replaces the Creator, or the emergence of biological novelties is attributed to an agent, the Creator, who replaces evolution.

## f. Irreducible Complexity

In *The Origin of Species* Darwin proposed how his own theory could be falsified. He wrote, "If it could be demonstrated that any complex organ existed, which could not possibly have been formed by numerous, successive, slight modifications, my theory would absolutely break down. But I can find out no such case."[52] We should not automatically assume that if Darwin could not find such organs they do not exist. Since Darwin, biology has made enormous progress in all fields and subfields. How well founded, therefore, is the nineteenth-century conviction of Darwin? And what if not one but multiple examples that cause the theory to "absolutely break down" have been found?

To answer these questions, first we need to formulate the problem more precisely. If we were talking about the emergence of complexity, Darwin could be right. Everywhere in nature we see complexity as well as order emerging spontaneously (the formation of crystals, complex structures of rocks exposed to erosion, waves in water, formation of clouds, etc.). The gradual (neo-)Darwinian mechanism could possibly produce complex organs because complexity can emerge gradually by adding, step by step, new parts to the whole. The problem, however, is that not all biological complexity is reducible to simplicity without losing functionality. And function is needed in order for natural selection to see the random modifications, because only function gives survival advantage. If a modification of an organ (or any biological structure) does not make the organ perform its function better (in terms of survival fitness), it will not be visible to natural selection. And here the problem begins.

52. *The Origin of Species*, 190.

In living organisms, there are some organs that can perform their function only when their total complexity is present at once, that is, when all their parts are, at the same moment, present and properly fixed in place.

Michael Behe, a biochemist who developed the idea of irreducible complexity in modern biology, refers to the example of the bacterial flagellum. A flagellum is something like an exterior propeller giving a bacterium the ability to move in liquids. The flagellum performs a function that is indispensable, or at least useful, for survival. The bacteria with nonfunctional flagella in natural conditions would be outcompeted by those with functional flagella. However, as studies have demonstrated, the removal of even one crucial protein from about fifty needed to build the flagellum will cause a complete loss of function.[53] All crucial proteins must be present in the flagellum to make it work, provide the survival advantage, and thus move evolution forward. The creation and proper assemblage of the fifty proteins in one evolutionary step (i.e., one generation) exceeds by far what can be developed by means of random variation. As Darwin himself acknowledged, his theory requires that the complex organ be produced by "numerous, successive, slight modifications." Darwin's mechanism might account for *complexity*, but it does not explain the origin of *irreducible complexity*.

Critics of Behe's argument refer to co-option as the way in which irreducibly complex systems could have come about. We have already discussed the logical problem of co-option (II,2,e). Now, using the example of a bacterial flagellum, we can demonstrate how the logical problem is confirmed in real biological conditions. The critics say that some elements of the flagellum exist in cells and perform different functions. However, they fail to mention that not all but only about half of the flagellar proteins are present elsewhere in the bacterium. None of the critics has shown a realistic biological scenario for how all of the proteins would come together, adjust to perform a new function, be supplied with the missing parts, and constitute a new functioning whole. Following Darwin, they concede that this cannot happen in one evolutionary step. Yet, if this were to happen in many steps, each of them would need to provide

---

53. Michael Behe, "Irreducible Complexity: Obstacle to Darwinian Evolution," 354; cf. *Darwin's Black Box*, 72–73. Some critics have indicated that the deletion of some proteins would not completely remove the function of the system. But this does not invalidate Behe's argument, because those non-essential proteins, by definition, do not belong to the irreducible system. Drawing on Behe's metaphor of a mousetrap, we can say that the non-essential proteins would be like color coating of the trap or a case to store it. Neither is necessary because neither contributes anything to the basic function.

not only functionality but also additional survival advantage over the previous steps. No one has come even close to showing realistic evolutionary steps toward producing the first flagellum.[54]

Moreover, to produce the flagellum, there must be not only the design of the working flagellum but also the assembly instructions, that is, the procedure for building a flagellum in every new bacterium. And this demands another design, which we do actually find in bacteria. They cleverly manage the building process beginning with production of the bushings in the cellular wall, then the engine, and the hook, and only at the end the flagellum itself. Hence, there are two designs for one irreducibly complex organelle that would need to come together in one evolutionary step.

Irreducibly complex organs demonstrate the inability of the neo-Darwinian mechanism to produce all complexity in biology. At the same time, irreducible complexity is a positive argument for design: the neo-Darwinian mechanism cannot produce functional novelty, whereas effects of this kind are known to be produced by intelligent agents who have the ability to act for an end. Contemporary microbiology is surprised by the fact that at the molecular level of biology we find devices that are similar to human technological developments—electric motors, gears and levers, suction pumps, advanced transportation systems, DNA codes and programs, etc. Intelligence is the only cause known to create irreducible complexity, because this demands foresight—the ability to assemble multiple parts which by themselves are not directly related to the function of the whole. The minimal conclusion is that at least some parts of living organisms must have been designed.

## g. Evolutionary Steps

In the second section of this chapter (especially II,2,d), we proposed a few logical problems with the transitions from one functional organ to another in a supposed evolutionary scenario. These problems are

---

54. Michael Behe describes this problem thoroughly in his books. In *Darwin's Black Box* he recounts a few proposals for how a step-by-step process could have produced the bacterial flagellum. However, those few proposals do not resolve even basic biochemical obstacles that would stop the process if it were to really happen in nature. In *Darwin Devolves* Behe shows that even new functional proteins are beyond the capacity of the Darwinian mechanism. See *Darwin's Black Box*, 65–73, and *Darwin Devolves*, 213–18.

# Evolution and Natural Knowledge

confirmed when we turn to actual biological structures. The vast majority of scientific literature in evolutionary biochemistry consists of gene sequence comparisons. Based on measurable similarity between genes or proteins, authors of academic papers draw conclusions about the degree of evolutionary kinship between proteins. But as we observed before, this conclusion is valid only under the *assumption* that the organisms carrying these proteins share common biological ancestry. Hence, the logic of the argument is circular.

Scientists connected with intelligent design have turned the issue around and asked the methodologically more correct question, namely, how much biological novelty can the neo-Darwinian mechanism actually produce? If the amount is satisfactory, the conclusion about common ancestry may be correct, but if the neo-Darwinian mechanism is unable to make the small steps from one working gene or protein to another, then the conclusion whereby it actually produced all genes or proteins is unfounded. Over the last decade or so, a number of experiments sought to estimate the actual power of random variations and natural selection; none of them provided much support for the neo-Darwinian mechanism of evolution.[55]

If the neo-Darwinian mechanism produces biological novelty by moving from one organic structure to another, it should be capable of producing new proteins quite easily. Proteins, after all, are the basic building blocks of living organisms, and thus obtaining any new organism would require many new proteins in the first step. We cannot use in an experiment a protein that does not exist, i.e., has not been "invented" yet by natural selection (which would actually be a real-case scenario). However, we can take two proteins whose structures are very similar but whose functions are different enough to be visible to natural selection. Let us reiterate here one of the experiments that, in our opinion, clearly shows the problem.

Scientists who performed the experiment[56] chose proteins called $Kbl_2$ and $BioF_2$ and determined which genes code for them. Then they established how many and what kind of changes in the genetic information are required to change $Kbl_2$ in such a way that it would perform the function of $BioF_2$. (Keep in mind that this is not a complete transformation of one protein into another. It is just altering one protein to

---

55. See, for example, experimental results of Douglas Axe published in the *Journal of Molecular Biology* and in *BIO-Complexity*, listed in the Bibliography.

56. Axe and Gauger, "The Evolutionary Accessibility of New Enzyme Functions."

a sufficient degree as to enable it to perform the function of the other. This should be quite easy for the Darwinian mechanism.) The relevant step in Darwinian evolution cannot be shorter than this, because as long as $Kbl_2$ does not perform any new function, no changes are visible to natural selection. It turns out that to make $Kbl_2$ perform the function of $BioF_2$, a minimum of seven nucleotides in the gene need to be replaced. The problem is that single mutations are frequent, double mutations are rare, and triple specific mutations do not happen due to extremely low probability. The probability of this event (seven nucleotide replacements to create the new protein) is inconceivably small; it was estimated that it would require $10^{30}$ bacterial generations (each generation is like a new attempt) to occur. Even though bacteria multiply fast, the time required for this number of generations vastly exceeds the history of life on earth.[57]

The problem with "evolutionary steps" reoccurs whenever scientists inquire into the ability of the Darwinian mechanism to produce biological novelty.[58] Biochemist Douglas Axe presents the problem in more general terms using the imagery of a robot searching through a rugged landscape. We quote his explanation at length:

> Darwinian evolution is often thought of in terms of journeys over a vast rugged landscape. Each point of this strange terrain represents a possible genome sequence, those possibilities being so staggeringly numerous that real organisms have only actualized a minute fraction of them. The ground elevation at each point corresponds to the fitness of individuals carrying that genome, with the horizontal distance between any two points indicating the degree to which the corresponding genomes differ. In terms of this picture, all of the millions of species alive today are represented by their own points, high up on peaks scattered

---

57. In the "Discussion" the authors write, "We estimate that some $10^{30}$ or more generations would elapse before a *bioF*-like innovation that is paralogous to *kbl* could become established. This places the innovation well beyond what can be expected within the time that life has existed on earth, under favorable assumptions. This places the innovation well beyond what can be expected within the time that life has existed on earth, under favorable assumptions." Axe and Gauger, "The Evolutionary Accessibility of New Enzyme Functions," 12. In fact, even the unrealistically favorable assumption that *kbl* duplicates carry no fitness cost leaves the conversion just beyond the limits of feasibility". p 12

58. On the limited possibility of creating new proteins by point mutations, see the study of Behe and Snoke, "Simulating evolution by gene duplication," listed in the Bibliography.

somewhere across this conceptual landscape (the fact that they are alive demonstrates the quality of their genomes).

Now, wherever a species happens to be, Darwin's engine tends to move it toward the highest ground it can reach [see Figure 2].

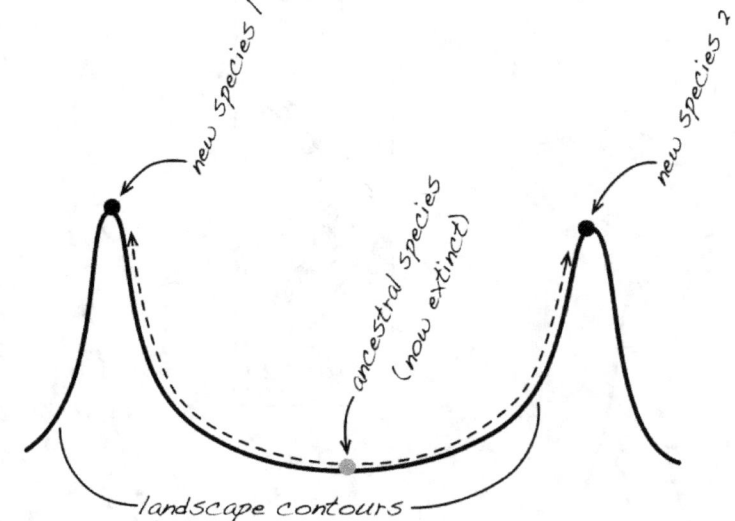

Figure 2. Darwin's explanation of the diversity of life forms. This is a cross-section through two peaks representing substantially different forms of life. The whole landscape stretches out in all directions, with millions of peaks representing all the different species.

According to the Darwinian story, that simple tendency to migrate upward has, over billions of years, transported the first primitive genome from its starting point to higher points along millions of diverging paths. The result is the spectacular variety of life forms we see today with a correspondingly wide dispersal of genomes across the vast conceptual landscape.

But there is something suspicious about this story . . . It has to do with the wide disparity of distance scales. The scale of the landscape, which is characterized by the extent to which dissimilar genomes differ, is very large by any reasonable calculation. On the other hand, Darwin's engine moves in steps that can only reach points a tiny distance away from the prior point. In one step it can move a genome to the highest within this reach, but further progress would require a still higher point to fall within reach once that move is made. That might happen every now and then, but it would have to happen in an amazingly consistent and helpful way to explain how the enormous distances were traversed from the point marking the first primitive organism to the millions of points marking the great variety of modern life forms. [See Figure 3.]

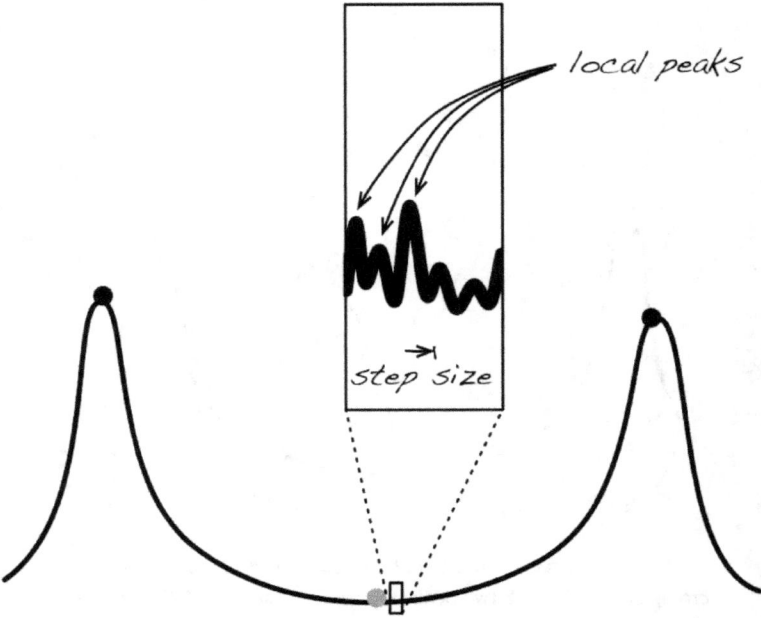

Figure 3. The problem of climbing in tiny steps. If the engine moves to the highest point that can be reached in each step and the landscape is rugged, then the endpoint will be a local peak.

Let's put this in more familiar terms. The summit of Mount Whitney, the highest point in the contiguous United States, is just 136 kilometers from the lowest point in North America, known as Badwater Basin. Now, suppose there were an automated vehicle capable of remotely scanning to the highest point identified by the scan. If the scan radius is greater than 136 kilometers, this vehicle could get from Badwater to Whitney in one scan-and-move operation. But what if the scan radius is one *millionth* that size? Now the circle that the vehicle "sees" from its current position is about a shoe-length across, with each move being up to half that distance. Considering how uneven the ground is, we wouldn't expect this nearsighted vehicle to complete more than a few scan-and-move operations before becoming stuck on a rock, maybe half a pace from where it started. Summiting Whitney would be completely out of the question. So the idea that *any* ability to seek higher ground, no matter how restricted, makes the highest summit accessible turns out to be highly simplistic.[59]

With this metaphor, Axe shows both the inability of the neo-Darwinian mechanism to create biological novelty and the inability of natural selection to make any significant steps toward new organs and new forms of life. By the way, we also learn from these experiments how science, if properly done, supports the conclusions of the sound philosophy that we presented above (in II,3). After all, truth cannot contradict truth.

### h. Is Darwinism More Scientific Than Intelligent Design?

Right from the beginning, the Wikipedia entry informs readers that intelligent design is "pseudoscientific," a "religious argument ... [that] lacks empirical support," and a "form of creationism."[60] None of this follows from what we have said thus far. Therefore, a good question to ask is the following: Why would a popular online resource label the theory with three apparently misleading descriptions? The most probable answer is that labeling it "creationism" or "pseudoscience" is the easiest way to dismiss the arguments for intelligent design without even attempting to provide a scholarly rebuttal. The Wikipedia entry is a good example of the misrepresentation that intelligent design theory often receives from popular media and books. And this is a good igniter for another discussion: What is the epistemological status of the theory? Regardless of the ideological baggage accompanying the discussions on intelligent design, the question of whether it is theology (as Wikipedia suggests), philosophy, or science is a legitimate scholarly problem.

To answer this question, we first need to know what science is. In the first chapter (I,1,a), we defined science as natural knowledge about the material universe. Natural knowledge means knowledge acquired by studying natural phenomena, without the help of supernatural revelation. There are dozens of different definitions of science. Nevertheless, our definition differentiates science from other forms of knowledge and thus it is useful when it comes to establishing the epistemological

---

59. Axe, "Darwin's Little Engine That Couldn't," 31–43, 36–39.
60. https://en.wikipedia.org/wiki/Intelligent_design.

status of ID in the context of philosophy and theology. From what we said about intelligent design (II,5), it follows that this theory is derived entirely from studying natural phenomena. It relies on methods typical of scientific arguments. It does not resort to any supernatural knowledge, such as sacred books or private revelations. It does not offer any "theory of creation"; indeed, it works quite well even without any idea of creation, let alone a Creator. Hence, saying that intelligent design is theology or creationism is simply misleading.

Still, if ID is not theology, perhaps it is philosophy, and for this reason it should not be taught in science classes? Philosophy (according to our definition in I,1,a) is a study of being by means of natural reason, without reference to revelation. Both intelligent design and philosophy study being without resorting to revelation. But there are at least three essential differences between the two.

First, ID is concerned entirely with the natural world, and, strictly speaking, just with biological structures. In contrast, philosophy is concerned with any kind of being, including real, immaterial (spirits) and ideal (ideas in the intellect). Second, ID is focused on particulars, whereas philosophy approaches reality in the most general and abstract way. In fact, this is precisely what philosophy means—the love of wisdom, and wisdom is the knowledge of ultimate causes. Hence, philosophy ultimately must be concerned with the first cause of everything, whereas intelligent design deals with causes of some material beings without even asking whether these causes are ultimate, proximate, or secondary. The third difference is in the method. Philosophy uses the natural common experience of the universe and first principles of cognition (*prima principia cognitionis*) in order to draw highly abstract conclusions about being as such. ID employs inductive and abductive modes of argumentation, probability theory, statistics, and experiment—all of which are commonly recognized methods in natural sciences. These three differences show that intelligent design, strictly speaking, is not philosophy.

However, for centuries, before natural sciences developed and grew into their own domains, science was considered just a part of philosophy—philosophy of nature. Natural philosophers were those whom we now call scientists. In this sense, ID as well as many other parts of natural science can be called philosophy. Calling any part of science "a philosophy" should not be considered derogatory, as if it diminishes the cognitive value of the discipline.

We also need to distinguish between a theory in natural science and the possible philosophical or ethical conclusions that the theory may suggest. A good example of confusion between these two is found in the history of Big Bang theory.

When, in the early twentieth century, the stationary model of the universe was replaced by Big Bang cosmology, some scientists opposed it because the new cosmology seemed to necessitate the Christian idea of *creatio ex nihilo*.[61] If the universe had a beginning, it must have been caused by something outside of the universe, which means that our physical reality is not everything that exists. And this was unacceptable for scientists of the atheistic mindset. However, any theological or philosophical conclusions that follow from Big Bang cosmology do not diminish the scientific value of Big Bang theory itself. Similarly, intelligent design provokes enormous resistance from the atheistic and materialistic portion of the scientific community, because if some structures in biology are purposefully designed, the designing intelligence must have existed *before* human intelligence came about. And this turns the grand materialistic-evolutionary story upside down. Yet, the impact on our worldview that the theory of intelligent design may have does not diminish the scientific validity of the theory itself. After all, it is possible to think about intelligently designed structures in nature without asking philosophical and theological questions regarding the nature of the intelligent cause and how the design was introduced into biology.

Finally, we need to acknowledge that many ideas—and especially theories of origins—consist of different layers of discourse and different levels of human knowledge. Hence, in some respects intelligent design is philosophy, and in others it is science. However, exactly the same applies to Darwinism—it includes science, but also philosophy or even theology, or rather anti-theology. The important question is not whether ID is science or not, but rather is ID on the same epistemological level as Darwinism? If they are, they can compete within the same domain by proposing alternative answers to the same questions. Apparently, if we consider the scientific content of ID, it can be juxtaposed with the scientific content of Darwinism. If we consider the philosophical implications of ID, they can

---

61. Even as recently as 1989, the editor of *Nature*, John Maddox, published an editorial accusing Big Bang cosmology of being philosophically unacceptable because of the religious ideas it implied. The author believed that the theory would not survive another ten years. He believed that a new solution, one that did not require an absolute beginning, would be discovered. See Maddox, "Down with the Big Bang," 425.

be contrasted with the philosophical conclusions implied by Darwinism. Darwin made it clear that the main goal of his theory of natural selection was theological in nature—he intended to remove the "repetitive acts of creation" from the natural history of the universe. Darwin did not stick to the purely scientific framework. Therefore, if ID is theological, that is even more the case with Darwinism.

Those who challenge ID from a naturalistic perspective say something like this: "ID is not scientific, because it does not provide any mechanism of *how* design was introduced into biology. In contrast, Darwinism is scientific, because it shows *how* biological structures emerge."[62]

A short answer to this charge is that if species were formed supernaturally (which is the classic Christian position), then there cannot be any material mechanism that would explain their origin. In this case, ID as a scientific theory would be perfectly compatible with the theological doctrine of creation. In contrast, Darwinism, by the very fact of providing a mechanism, excludes supernatural causation and thus becomes a reductionist philosophy. Hence, the fact that ID does not provide a mechanism may actually make it more scientific than Darwinism. Diagram 15 presents this problem in greater detail.

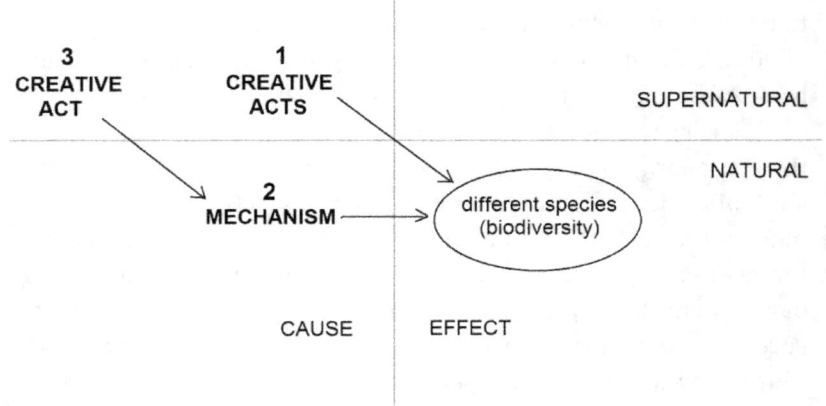

Diagram 15. Three types of divine causality in the production of species.

---

62. See, for example, Asher, *Evolution and Belief*, 32. Interestingly, Thomistic evolutionists do not show any mechanism in ID, but they charge ID of adopting a mechanistic view of nature. See also Beckwith, "Intelligent Design, Thomas Aquinas, and the Ubiquity of Final Causes," 6, and Edward Feser, "The Trouble with William Paley."

In Diagram 15, we see that one effect (the existence of different species) theoretically may be caused in three different ways. The first is through direct divine acts (1). This is the traditional Christian position derived from the Bible and usually called creationism (though not quite properly, see III,1,c). The second is through natural causation alone, such as we find in the Darwinian mechanism of random variation and natural selection (2). This is an atheistic approach. The third way of bringing species into existence is supernatural causation operating through the mediation of secondary natural causes (3). According to this position, God created different species using evolutionary mechanisms. This third position is called theistic evolution.

Now, some scholars believe that Darwinism favors primarily the second type of causality (atheistic evolution) but is also compatible with the third one (theistic evolution). Consequently, "theistic evolutionists" believe that Darwinism is reconcilable with Christianity. In fact, however, a vast majority of scientists use the term (neo-)Darwinism in a way that excludes any reference to divinity. According to their attitude, the Darwinian theory of biological macroevolution excludes not only the direct but also the indirect type of divine causation. Regardless of which party is right (i.e., theistic or atheistic evolutionists), neither of them accepts *direct supernatural causality*. In this respect, both parties follow Darwin, who was quite explicit in his desire to remove the direct supernatural causality from natural history. In contrast, intelligent design is compatible with direct supernatural causality (1) as well as mediated divine causality (3). By its very nature, ID cannot be compatible with the second type of causality (i.e., mechanisms alone), because neither laws of nature nor any natural mechanism can produce the purposeful design or specified information that we find in biology. A mechanism can only pass on a design; it can never create the design. For example, the design of the bacterial flagellum is passed on and multiplied through the natural process of generation and propagation among bacteria that possess the flagellum. But the same process does not account for the origin of the flagellum.

Now, the objection quoted against ID states that ID is not science because it does not offer any mechanism (of the emergence of species). ID does not offer it, but neither does it exclude it. But the more important observation is that a mechanism is possible only under the assumption that species were not created by direct supernatural causation. Yet, this assumption is not provable. If there *was* direct supernatural causation in the production of species, then ID properly delineates the limits of science. By the very fact that it does not provide a mechanism, it stops before it goes too far. In contrast, Darwinism addresses the question of the origin of species in such a way as to enter the field of theology by excluding the direct divine causation. It thus appropriates some area of knowledge that does not belong to science (cf. Diagram 7B). If species were actually formed by supernatural and direct creative acts, then Darwinism must end up as either an alternative religion or materialistic reductionism. In any event, it seems that a scientific theory that recognizes the limits of science is more scientific than a theory that extends beyond the domain of science.

## i. The Limits of Methodological Naturalism

The principle of methodological naturalism says that when we want to explain a phenomenon, we should resort to natural causes only. Since science deals with natural phenomena, this principle seems to be justified within the natural science. The problem begins when the advocates of methodological naturalism propose that *all* phenomena must have a natural explanation. And what if a phenomenon does not have a natural explanation, as is the case with the miracles? Then the natural explanation cannot be true, and the advocates of methodological naturalism are doomed to defend a wrong, reductionist explanation. For this reason, the absolute commitment to the principle of methodological naturalism is not the proper approach for a reasonable person who wants to know true answers to important questions. Apparently, a better approach is to look for *the best* rather than *natural* explanations, without establishing *a priori* which explanations are the best. As a popular saying goes: "Scientists should go wherever the evidence takes them." After all, it should not be surprising, at least not for a Christian, that some phenomena in the natural world do not have a natural explanation. It seems quite probable that God would have created the universe in such a way as to leave some gaps in the essentially uniform physical order, so that man can see Him through these gaps.

Still, the question remains whether or not science can ever violate methodological naturalism. As human beings, we should search for the true or the best rather than natural explanations. Scientists, however, deal with nature, and their task is to find natural explanations. Therefore, their commitment to methodological naturalism is justified; in science, the principle of methodological naturalism should never be abandoned. Otherwise, scientists might enter other domains of knowledge, such as theology and philosophy, thereby straying from their pursuit of the scientific truth. This would lead to confusion of methods, discourses, and conclusions, to the detriment of all domains of knowledge. There is, however, a condition that applies to this principle: methodological naturalism is valid only as far as science goes. But there are insurmountable limits of science; science cannot explain everything. And there are at least three areas in which science should recognize its limitations:

1. God and the immaterial world of spirits;
2. miracles, i.e., supernatural events in the material world;
3. questions of origins, specifically the origin of the universe, of human beings regarding their souls and bodies, and the origin of species.

Now, different scientists have different opinions about the limits of science. Atheists and materialists do not care about the first limit, because they do not believe that there is anything but the material universe. Scientists who are Christians usually accept the first and second limitation. But when it comes to the question of origins, there is great division among Christians. Some would say that the origin of species is a purely scientific problem (interestingly enough, there are also theologians who say so). Others would say that science needs *some* philosophy and theology. But very few would accept that science simply *cannot* say how species came into existence.

This rejection of the third limit of science often stems from misunderstanding the issue. Even if the origin of species is beyond science, it does not follow that science cannot speak about the history of species. Indeed, disciplines, such as paleontology, are valid science when they study the history of species—when they emerged, how many there were during this or that era, how long they lasted, whether they migrated or not, whether they adapted to different climates or not, and so forth. But the *origin* of species is a different type of question—it refers to the origin of entirely new and complete natures. Such a thing as a *complete new nature* cannot emerge little by little, step by step. It must be brought about all at once, in one act. And the question of origins concerns these acts that bring about new species. Since this kind of acts cannot be natural, they cannot be explained by science. Thus, the third limit of science, which is also the limit of methodological naturalism, is as valid as the first two.

In our discussion of the Book of the Bible and the Book of Nature (I,1,c) we said that science speaks about supernatural causation only in a negative way. Supernatural causation is the one that produces species (among other things). But science cannot investigate supernatural causation; therefore, science can investigate everything material that happened right before and right after a new species was created, but cannot provide a positive explanation of what happened between these two moments. It can, however, demonstrate that any scientific (i.e., natural) explanation is inadequate and fails to explain the appearance of the new species. This is what we call negative knowledge—science says that "this explanation proposed in science does not account for the origin of species." If scientists say that one of the proposed natural theories explains the origin of species, this is the moment when the limits of science and also the limits of methodological naturalism are violated. We see that this limit of science is confirmed in the scientific critique of neo-Darwinism proposed by the ID theorists. Much of their work is showing how different naturalistic explanations do not really explain the origin of species or of some essential structures required to build new forms of life.

However, ID theorists propose also a positive part of their enterprise. It consists of establishing an intelligent cause as the explanation for the origin of species. They do not propose any material mechanism to account for the phenomenon of growing biological complexity that we observe in history of nature. We said in a previous section (II,5,g) that by not providing a mechanism, ID may be actually more scientific than Darwinism, because it stops where it encounters the limits of science. Proposing that a mechanism (i.e., a set of material causes working together toward one goal) explains the origin of species is a transgression of the limits of science, because such a proposal tacitly replaces supernatural divine work with the work of nature. Clearly, there is no such reductionism in saying that intelligent design caused species to arise. So the problem of extending methodological naturalism beyond its scope does not occur in the theory of intelligent design.[63]

There is, however, an opposite charge that is often filed against ID, namely, that ID introduces supernatural causation into the domain of science, thus violating methodological naturalism. This kind of violation

---

63. Of course, the scientific statement that intelligent design causes species to arise can be given a religious interpretation, i.e., that intelligent design comes from God who educed forms from matter by His direct causality. But that religious interpretation is not required of ID theorists, and is not part of ID per se.

would be a valid charge against the scientific status of ID if ID were proposing supernatural causation where natural causes are satisfactory. If this were the case, ID would extend theology into the realm of science, falling into the opposite error from the one found in Darwinism (which is extending science into the realm of theology). In order to answer to this charge, we need to clarify a few terms and introduce some distinctions.

## j. Does Intelligent Design Violate Methodological Naturalism?

We defined the principle of methodological naturalism as a search for natural causes of material phenomena. The first thing to notice is that "natural" does not equal "material" in a strict sense. The very notions of "matter" and "material cause" are very difficult to define in modern science. Is gravity a material cause? If yes, then we have no clue what matter is. If not, then assuming that science can appeal to material causes only, gravity would not be a scientific explanation, which clearly is not the case. Is radiation a material cause? There are some particles involved in generating radiation, but radiation itself is a wave. Moreover, radiation is typically considered the opposite to matter, so it cannot be called a material cause in the strict sense.

Another difficulty in defining strictly material causes as the only explanation in science is that many scientific disciplines resort to mathematics or theoretical modeling, yet these things do not exist in matter. Most scientists agree that besides matter there is some virtual inter-subjective sphere that any person can potentially appeal to, that consists of at least logics and mathematics (something akin to the Teilhardian "noosphere"). Even those who do not believe in the immaterial nature of the human mind usually agree that mathematics does not exist in the same material way as physical objects around us. This sphere, unlike physical objects, cannot be observed (not even indirectly); it escapes any measurement or detection by senses. At the same time none of the given examples (gravity, radiation, the "noosphere") is considered a supernatural cause in modern science.

Therefore, material cause in science does not equal causation by material objects. In science material causation is defined somewhat broader, as any causation that can be studied through experimental method. For example, we do not see gravity, but we can study the effects of this force and recently we even discovered gravitational waves. The same applies to radiation—for the most part it cannot be watched, touched or sensed, but it is still considered material cause, because it can be measured and put under the test using scientific equipment. Hence, we can say that *science does not look for a material cause strictly speaking, but rather for a cause that can be measured, tested and described by the scientific method.*

The second thing to notice is that science does not look for all four Aristotelian causes of physical structures and events (for the meaning of the four causes, see II,3,d). The material cause for Aristotle is pure matter (Lat. *materia prima*) which is pure potency—an intellectual concept that may exist in reality only in combination with the form. Hence, material cause for Aristotle means something different than material cause for modern science. The four Aristotelian causes belong to the philosophical level of explanation of the physical being (natural philosophy). But the scientific level of knowledge is different from philosophy and one of the consequences is that science does not look for the four causes.

Applying Aristotle's scheme of four causes to modern science, we can say that science is concerned neither with the material cause nor the formal cause. Instead, modern science looks primarily for the efficient cause, and sometimes (secondarily) for the final cause.

The final cause can be understood in two ways. One way is to speak about the ultimate end of a particular being or the entire universe. In this sense we can say that the end of animals and plants is to provide for man, the end of the universe is to secure space for human prosperity, and the end of humans is to reach the unity with God in heaven. These ends can be called transcendent final causes. The second way of understanding the final cause is limited to the proper end of each material being. Thus, for example, the goal of a bird is to preserve its life, to nest, to lay eggs in order to propagate and preserve its species. The goal of a heart is to pump blood in order to distribute oxygen and nutrition to the body, the goal of gravity is to maintain cosmic structures and movements, etc. We can call this type of finality immanent. Science is concerned only with the immanent type of final causality.

The efficient cause, is the one that causes things to happen. We can compare the efficient cause to a "pusher" that causes things to move. (In contrast, the final cause is like an "attractor"—we can compare it to a magnet that causes an object to move in a specified direction.) In science, the efficient cause may be identified with the material cause (material in the scientific sense explained above). However, the essential meaning of the efficient cause in science does not refer to the nature of the cause (because, as we said, it is hard to define), but to the fact that the efficient cause is accountable for in scientific terms—it can be somehow observed, measured, put under the test, etc. Therefore, we can say that *science looks for the efficient and for the immanent final causes only.*

It is not reductionism when science focuses only on these two causes, because the other two go beyond its scope; they belong to philosophy. Reductionism takes place not when a scientist says that science is limited to these two, but when a scientist (or whoever else) claims that the other two causes do not exist absolutely speaking, or that there is no other form of knowledge than natural science. If scientists are aware of this limitation of science, they do not fall into reductionism by limiting their research to just the two types of causes.

Now, the question is whether intelligent causation postulated in the theory of intelligent design can be considered the type of efficient or final cause that remains within the limits of science. If so, then ID does not break methodological naturalism; if not, then it does.

In order to answer this question, we first need to expound on the nature of intelligent cause. We already established that ID does not make any statements regarding the nature of intelligence acting as a cause in biology. For materialists, intelligence is just a function of highly organized matter (the brain), which is not typically considered an immaterial cause, unless in the scientific sense defined above, that is, the same sense we call gravity and radiation immaterial. (*Nota bene*, there is a little bit of irony in the fact that materialists, who do not consider human intelligence any immaterial or supernatural force, would be the first to claim that the intelligent cause postulated in intelligent design is a supernatural cause that contradicts methodological naturalism.) Classical philosophy and Christian faith deem intelligence immaterial, albeit each of them for a different reason. But the understanding of the immateriality in philosophy and religion is different from the one adopted in science. In philosophy, intelligence is *immaterial* not just because it cannot be touched or sensed in any way, but because it is *spiritual*. Philosophy provides a positive explanation of the nature of intelligence—it is immaterial because it constitutes a different type of substance, which is characterized by (among others) immateriality, consciousness, rationality, freedom, and personality.

As we said, the goal of science is to find the efficient cause, but it is not essential what the nature of this cause is. Perhaps some scientists would gladly eliminate non-material causes, but as we noticed, there are many causes accepted as scientific that are hard to classify as material causes in a strict sense. It seems, therefore, that an intelligent cause can also be classified as one of them. ID does not postulate anything about the nature of intelligence, and similarly, science cannot explain the nature of some causes that are commonly recognized as scientific. It follows that it is not the nature of the cause (whether it is material or immaterial) that is crucial for science, but what it does (is it an efficient cause or not?) and whether it can be accounted for by the scientific method or not. The foundational claim of ID theorists is that the intelligent cause can be detected by the scientific method (see II,5) and that it is the actual efficient cause of some structures and events in biology. If they are right, then it is proper to accept intelligence as a legitimate cause in science.

For example, we do not know of any other cause but intelligence that can produce irreducibly complex biochemical systems. One reason is that to produce such systems the cause needs to have foresight—the ability to predict the outcome before it occurs—and intelligence is the only cause known to have this. Another reason is the fact that DNA contains specified information. Since information is detectable by the scientific method and intelligence is the only cause known to produce information, intelligence should be incorporated into science as the explanatory factor, i.e., the efficient cause for the existence of biological information. This does not break methodological naturalism, because even if intelligence is spiritual (as sound philosophy and Christianity maintain) the effects of its workings are accountable for by the scientific method (as they are in the case of gravity). As we said, not the nature of the efficient cause, but its accountability in scientific terms, is what matters in science. And since intelligent causation can be detected by the scientific method, it remains within the scope of science. Therefore, intelligent design does not break the principle of methodological naturalism.

Again, we need to remember that what a given scientific explanation means *per se* is one thing, and what kind of philosophical or religious conclusions one can derive from the scientific explanation is another. The Big Bang suggests creation out of nothing in the beginning of time. Even so, it is commonly recognized as a scientific theory. Intelligent design suggests the existence of a superior immaterial and intelligent Creator, but this philosophical conclusion does not make the theory of ID less scientific.

We should also notice that according to sound philosophy and Christian faith an intelligent being is not just immaterial i.e., non-physical, but also spiritual. Spirituality is a positive explanation of the kind of substance intelligence is. Besides, it is transcendent, which means that it is not limited to three dimensions (or any more or less than this), or any space, or place, or any Aristotelian category. Also, the causation of an intelligent being needs to be supernatural, because intelligence does not belong to nature, i.e., the physical universe. Nevertheless, these are all philosophical conclusions not possible to achieve within science. Natural science accepts intelligent causation not because it is spiritual, transcendent or supernatural, but because it is intelligent, and it is the only type of cause that can produce some of the effects found in nature.

### k. Methodological and Ontological Naturalism

Methodological naturalism in science says that we should never resort to supernatural causation in scientific investigation. The proponents of theistic evolution defend methodological naturalism against intelligent design in science and against the traditional understanding of creation in theology. They are, however, aware of the subtle problem in their position—namely, that if methodological naturalism is employed beyond the scope of science, it ends up as materialistic reductionism, i.e., materialism. To avoid an implicit acceptance of such a blatantly anti-Christian position, theistic evolutionists distinguish between methodological (or epistemological) naturalism and ontological (or metaphysical) naturalism. The former states that we should only resort to natural explanations, and the latter that nothing exists other than the material universe. After presenting this distinction, theistic evolutionists conclude that Christians accept methodological naturalism but reject ontological naturalism. According to theistic evolutionists, the rejection of intelligent design in science does not make them materialists, because it does not imply adoption of the ontological type of naturalism.

As much as this solution may look coherent, it misses some important details, which we can provide, borrowing from the discussion on naturalism in Chapter I and the previous subsection. First, we agree that methodological naturalism differs from ontological naturalism. Second, we agree that methodological naturalism should be maintained in science even if it is not applicable in philosophy or theology. Third, unlike theistic evolutionists, we believe that questions of origins exceed the competence of science, and thus that when one seeks the answer to them, methodological naturalism must be suspended. Fourth, we do not think that the theory of intelligent design necessarily violates methodological naturalism, as it simply does not say whether the intelligent cause is natural or supernatural (as has been shown in the previous subsection). This kind of judgment belongs to philosophy and theology. If intelligent causation can be detected by means of the scientific method, natural science should incorporate it. Consequently, intelligent design can be included in science without violating the principle of methodological naturalism.

## Chapter II—Summary

In this second chapter, we have established that the theory of biological macroevolution encounters many difficulties, both in philosophy and in science. The very logic behind the key concepts, such as transformation of species and natural selection, is questionable. Also, classical metaphysics challenges the core claim of the theory, namely, that accidental changes in individuals can produce new substantial forms or living beings of new natures. Further, many biological facts, such as the fossil record and the nature of genetic mutations, do not support the neo-Darwinian type of biological macroevolution. In the light of these challenges, another theory needs to be proposed, with intelligent design being a strong candidate. As we have shown, ID is as scientific as (in some aspects even more scientific than) neo-Darwinism. However, it is not the scientific status that justifies the debate between neo-Darwinism and ID, but rather the fact that both theories are on the same level of knowledge: both incorporate strictly scientific and somewhat philosophical elements. Thus, we can conclude that a particular type of biological macroevolution, called neo-Darwinism, should be replaced with intelligent design because of the evidence coming from biology, and that the idea of biological macroevolution (regardless of the mechanism) should be abandoned altogether because of the evidence coming from philosophy. In the next chapter, we will see how these changes would bring us to a new science–faith synthesis.

# Chapter III

# Evolution and Christianity

HAVING PRESENTED THE RELATION between evolution and natural knowledge in Chapter II, in this chapter we will explain how evolution relates to the Christian faith. Our goal is to expose some difficulties that the theory of biological macroevolution faces in the light of Christian theology. First, we will outline the arena of the current debate. This should help us define what is at stake in the debate over evolution (Section 1). Then we will present the classic Christian concept of creation in order to show how it differs from the evolutionary theory of the origin of species (Section 2). Next, we will introduce a new concept of origins, called "theistic evolution," which originated in the nineteenth century and over the last century generally superseded the classic concept in most Christian denominations (Section 3). Following this, we will address the common opinion among theistic evolutionists that the theory of intelligent design is hostile to, or at least incompatible with, Christianity, and will show how this charge is mistaken (Section 4). After this, we will expound on a new model of the relation between faith and science that is possible only now, after the theory of intelligent design has been proposed in science (Section 5). Finally, we will justify the very possibility of challenging the prevalent cultural convictions from the religious standpoint (Section 6).

## 1. The Arena of the Current Debate

### a. The History of Western Thought— from Paganism to Neo-Paganism

Christianity substantially transformed the image of God in pagan culture. The Church taught that God was not a lonely monad, but a Trinity and a Communion of Persons. Christianity also revealed to humans the truth about man—who he is and what his origin and destination are. It introduced the proper understanding of the dignity and equality of human beings in both forms—male and female. Faith in Jesus Christ brought also a new understanding of the relation between God and man: man is not a mere toy in the hands of gods or blind godlike forces (such as the Greek *Ananke* or the Latin *Fatum*), but rather an intentionally created being who is loved by God and called to salvation and union with Him. From a broader perspective, we see that every aspect of human culture and belief was revolutionized by Christ and gained some new understanding in Christian light. This "revolution" includes the understanding of the origins of the physical universe as well. Christianity brought a new worldview into the Pagan culture.

Ancient philosophers, such as Plato and Aristotle, by means of observing nature and drawing more general conclusions about the causes of the universe, discovered the Supreme Being, the Absolute. They also considered the origins of the universe, but they never came up with the idea of the creation of the universe out of nothing or the absolute beginning of time. Instead, most philosophies (and mythologies) maintained that the physical universe is eternal. In pagan thinking, there was always something coexisting with God that was not God Himself. Therefore, we can say that the idea of the absolute beginning of the universe is specific to Judeo-Christian faith based on the Bible. Current scientific data, gathered under Big Bang theory, conform with this ancient biblical truth to a surprising degree.

But the contrast is more extensive than this. In pagan philosophies and mythologies, God was too great and too far removed from the universe to act *directly* in the physical order. Indeed, it was not the idea of *maintaining* the universe in existence but rather the belief in God's *direct* action in the world that immediately differentiated Christianity from pagan beliefs. We will not find the concept of direct causation by a transcendent God in any of the ancient philosophies or religions. For example, Aristotle taught that the first cause maintains the universe in existence, but he would never allow the first cause to act directly upon the lowest being, that is, the physical universe. Instead, for Aristotle, the first cause acts upon the first out of fifty celestial spheres—the one that is closest to the Supreme Being and at the same time furthest removed from man. Similarly, in the Platonic and Plotinian cosmological systems, the pre-One (the Supreme Being) emanates different spheres (enneads) that constitute different orders and hierarchies, but only the lowest of them constitute and animate the physical universe. Thus, here again, we find a greatly developed system of secondary causes, but no idea of God's direct operation in the lowest spheres of the universe.

In contrast, Christian teaching includes three essential novelties: (1) the first creation of the universe out of nothing; (2) its supernatural and direct formation during a period of time presented in Genesis as the six days; and (3) the direct creation of the first human out of dust of the earth. The creation of man completes the work of the supernatural formation of the universe; however, God continues to work supernaturally in the history of salvation through miracles, revelations, infusions of grace into human hearts, and many other supernatural activities and interventions.

Christianity did not exclude or downplay the role of secondary causes operating in the universe. In fact, from the very beginning, Christians affirmed the mediation of the saints and angels in our communication with God. Christian theologians, such as Thomas Aquinas, explained thoroughly that God can act in the universe by means of secondary natural or supernatural causes. Nevertheless, the true novelty of Christian faith and the great discovery of Christian theology was not secondary causation but primary and direct divine causation extending to the lowest and even most insignificant creatures. This kind of divine "transcendence" was not possible or imaginable for pagans. Moreover, Christianity not only proclaimed the *ability* of the first cause to act directly in the physical order, but affirmed that it actually had done so, in creation and frequently, afterward, in the history of salvation.

Over the centuries, Christian teaching about creation gradually gained acceptance, and eventually completely pervaded the European way of thinking about the origins of the universe. The biblical paradigm was firmly established in people's minds. Yet, in early modernity, a set of events, including religious and political turmoil and unexpected discoveries in geography and astronomy, undermined the faith and the worldview of medieval people. Western civilization started doubting its well-defined dogmas, and the processes of secularization crept into mainstream culture. It is not a coincidence that along with the abandonment of faith, new ideas about the origins of the universe emerged. The old pagan idea of secondary causation acting on behalf of God regained its influence and appeal. Enlightenment philosophers proposed that God worked only at the beginning of the universe and then left everything to secondary causes (this view was an important aspect of deism). Fascination with the unchangeable laws of nature discovered by science made secular philosophers believe that God was needed only as a keystone for philosophical ideas, not as the one actively present in history. Moreover, they postulated that God who creates by secondary causes is nobler, more sublime, as this mode of creation reveals more of His ingenuity. As an Anglican theologian of the day, William Paley, put it, "God prescribes limits to his power, that he may let in the exercise, and thereby exhibit demonstrations of his wisdom."[1] Similarly, Erasmus Darwin, the grandfather of Charles Darwin, was enchanted with the idea of God acting entirely through secondary causes: "What a magnificent idea of the infinite power of the Great Architect! The Cause of Causes! Parent of Parents! *Ens Entium!*"[2] Over the following two centuries, the idea of special creation was almost completely removed from the European worldview. The same process affected Christianity itself, which generally adopted the "new creation story" in the form of theistic evolution.

We see that the modern, and especially the positivistic, understanding of God's causality is similar to the pagan one, even if in modern times it is formulated in a different language—the language of advanced science. The theory of origins proposed by Charles Darwin came at a perfect time: the direct divine causation was already challenged by deists and its place in our understanding of God was downplayed by the physicotheology of Paley-like theologians. Additionally, the role of theology in explaining the universe was undermined by positivism—the idea that only natural and empirical knowledge is objective and worth pursuing. However, at the dawn of the nineteenth century, nobody knew yet what the natural explanation of the origin of species should look like. Darwin offered a "mechanism," and thereby provided what had been missing for long.

It does not matter whether the evolutionary mechanism was merely started by the Creator, was continually assisted by divine providential

---

1. Paley, *Natural Theology*, chapter 3, 26.

2. Erasmus Darwin, *Zoonomia*, iv, xxxix, 8, 509. For more on this matter, see my *Catholicism and Evolution*, 64–71.

care, or completely eliminated the need for a Creator. The core of Darwin's theory is the rejection of the direct divine causality which for centuries constituted the *differentia specifica* of the Christian worldview. And the more Christianity is removed from culture, the more Christian principles are challenged. Hence, in our times, theistic evolution replaced direct causation and reintroduced secondary causation, which was a hallmark of the ancient pagan worldview. Theistic evolutionists emphasize that God maintains the universe in existence, but they are reluctant to acknowledge God's activity in history, and also, little by little, some of them eliminate *creatio ex nihilo* and promote the multiverse hypothesis instead.[3]

Every civilization needs a foundational myth—some universal story that shapes a people's worldview. The foundational myth functions as the predominant conviction for mainstream culture and as a paradigm upon which the civilization grows. We can call it a common belief or a cultural code. For the pagan civilizations, this had two forms: one popular, for the broader audience, i.e., ancient mythology; the other, for the educated people, i.e., philosophy. In mythology, gods are usually mixed with creation to the point that it is hard to tell who is a god, who is just a hero, and where the supernatural ends and the natural begins; in philosophy, God or gods appear as distant from the physical universe and hardly ever acting directly in it.

In contrast, in Christian civilization, the foundational myth was the historical and realistic understanding of the Bible. The book of Genesis was considered not just one of the ancient stories, but an actual description of how God made the universe. In our times, the role of the foundational myth has been assigned to the general theory of evolution.

The foundational myth is always formulated in language that is authoritative and convincing for the people of its era. For ancient pagans, this was the language either of mythological gods or of remote divine first principles. For Christians, it is the language of historical realism in which God unveils His plan in history through actual events. These events bear witness to a constant encounter between the supernatural and the natural, the divine and the human, the eternal and the historical. For neo-pagan societies, the foundational myth is formulated in the language of science, for science is the main force governing modern culture. We see, therefore, a kind of circular movement in the history of human

---

3. A good example of such a view of evolution is found in the works of the Polish philosopher and priest M. Heller. See his *Filozofia przypadku*, 32, 197, 268n.

thought regarding the origins of the universe. Diagram 16 summarizes these mental shifts.

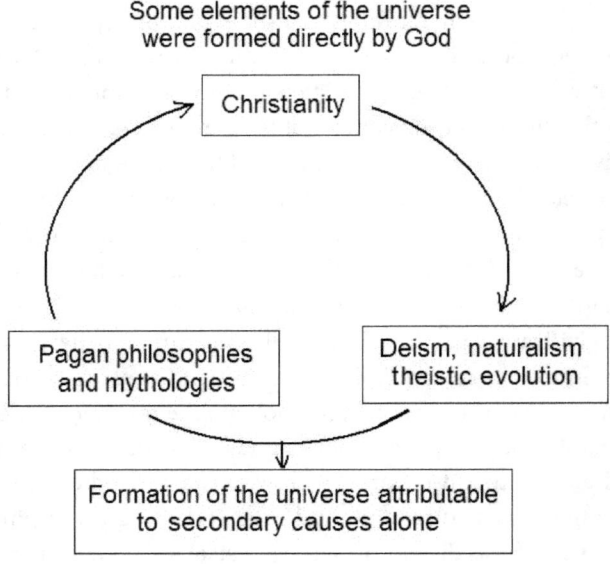

Diagram 16. The circular movement of Western thought
from paganism to modern evolutionism.

## b. The Four Concepts of the Origin of Species

Even though theistic evolution is the predominant view in our times, there are still some other ways of how people understand the origin of species in the current debate. The basic division includes these four: materialistic evolution, theistic evolution, progressive creation, and young-earth creationism. Let us briefly look at each of them.

1. *Materialistic evolution* (ME) says that the entire universe assumed its current form as a result of the laws of nature and inherent properties of matter, such as the imagined ability of matter to self-organize. Materialistic evolution includes *atheistic evolution*, but it is not identical with it because materialistic evolution does not require the nonexistence of God.

Materialistic evolution proposes that the universe is explainable without any direct or supernatural indirect causation or providence. Everything that is found in nature can be explained by the operation of physical laws. In this sense, materialistic evolution can be also called *agnostic evolution*, because it does not say anything about God—it does not exclude His existence, neither does it need Him to explain anything. Thomas H. Huxley coined the term "agnostic" to describe the religious position of the first proponents of biological macroevolution, such as Charles Darwin, Ernst Haeckel, and himself. The approach of these first evolutionists could be referred to as theoretical agnosticism and practical materialism. Haeckel advocated materialistic monism, a position in which spirit and matter are one, because spirit is just the higher organization of matter.

The atheistic form of evolution was fully developed only in the mid-twentieth century by authors such as Ernst Mayr, Julian Huxley, Jacques Monod, and George G. Simpson. For them, biological evolution entirely excludes the existence of God. Contemporary representatives of the atheistic type of evolution include Richard Dawkins and Jerry Coyne. The emergence of new natures in materialistic evolution is depicted in Diagram 17A. Here God is absent from the picture, and all natures are descendants of the previous natures that produce them through natural generation.

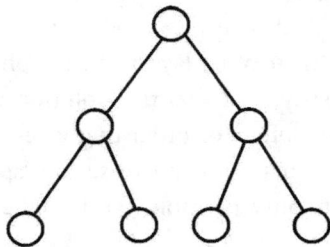

Diagram 17A. Materialistic Evolution. Circles represent new natures (such as natural species). Lines represent the cause–effect relations.

2. *Theistic evolution* (TE) states that God used evolution as a tool, or secondary cause, to bring about all living species. There are three basic forms of theistic evolution:

– The *deistic* form postulates that God set up the "initial conditions" at the first moment of creation in such a way that the universe developed toward higher forms thanks to the laws and properties inherent in physical reality. (This is the so-called "front-loaded" evolution.[4])

– The *pantheistic* form of theistic evolution says that God not only set up the grand evolutionary process in the beginning but also guides it over time. According to this concept, God is somehow present in evolution, either as a concurrent or a final cause. In some versions, God either co-evolves with the universe or must limit His power in order to allow physical laws to work according to the full capacity He gives them. Some proponents of the pantheistic form of theistic evolution accept that God is different from the universe, and some even defend His transcendent nature. Nevertheless, these authors do not clearly show how God works in evolution and how this differs from the work of nature alone, or how God is supposed to guide an entirely unguided process. This most popular form of theistic evolution leans toward and eventually ends up in pantheism.[5]

– The *emanationist* form says that the physical universe is governed by universal and eternal laws that are, in turn, governed by a higher order of laws, such as mathematical and logical principles, that ultimately come from God. The number, the character, and the role of the layers of reality may differ from one emanationist system to another. Sometimes God emanates the mathematical superstructure of the universe, and sometimes He is identified with this superstructure. The mathematical superstructure may emanate other layers, such as physical laws or logical rules. The emanationist type of theistic evolution implies that there is some permanent higher sphere or order that gives ultimate direction to random evolutionary events in the universe, and specifically in biological evolution. In emanationist cosmology this higher sphere is granted a divine status.[6]

There are a few points common to all three types of theistic evolution. One is that God does not exercise His direct or supernatural power in the

---

[4]. This is essentially Howard Van Till's idea of "creation's functional integrity." See his "Basil, Augustine, and the Doctrine of Creation's Functional Integrity."

[5]. This tendency appears in both Arthur Peacocke's "panentheism" and John Polkinghorne's "dual aspect monism." Even less "radical" authors clearly lean toward some form of pantheism, such as James Iverach with his "immanence of God in the universe." See Peacocke, "Welcoming the 'Disguised Friend'—Darwinism and Divinity," 471–86; Polkinghorne, *The Faith of a Physicist*, 21; Iverach, *Christianity and Evolution*.

[6]. The emanationist type is represented by e.g., Polish priest and philosopher, Michael Heller. See his *Filozofia przypadku* [*Philosophy of Chance*], 313n.

order of creation. Some theistic evolutionists would go further, claiming that God is unable to work directly,[7] but the majority are content with the idea that God simply chose not to act supernaturally in the formation of the universe, leaving everything to natural secondary causation. Some theistic evolutionists also reject the idea of miracles;[8] however, most of them acknowledge that in the history of salvation (after creation was completed) God occasionally worked supernaturally upon nature.

Second, most theistic evolutionists acknowledge that the universe was created out of nothing at the beginning of time, but affirm or imply that the formation of the universe brought about by evolution is not yet finished; for in theistic evolution entirely new species of living beings can still emerge, and thus creation is not completed. God continuously creates the universe (*creatio continua*) or supervises the self-formation of the physical order.

Third, theistic evolutionists usually say that design in nature cannot be detected by science. Instead, studying design and teleology in nature belongs exclusively to philosophy and theology. Many theistic evolutionists agree that Darwinian mechanisms may be unsatisfactory for explaining all the diversity of life, but this does not undermine the very idea of

---

7. This way of thinking appears in process theology as presented by Alfred N. Whitehead and his followers (Charles Hartshorne, Bernard Loomer, et al.). According to Whitehead, God evolves along with the evolving universe. Thus, he does not command or absolutely control events in the universe; rather, the events (the world) make impressions on God, molding him accordingly. See *Process and Reality* 346–51. It is hard to avoid an impression that even John Polkinghorne's idea that "creation is an act of divine kenosis, an expression of the self-limitation" tends toward limiting divine omnipotence. See Polkinghorne, *Science and Creation*, preface.

8. Typically, the challenge to miracles is not formulated explicitly (because this would flatly contradict Christian doctrine), but only implied. For instance, an article on miracles at the BioLogos website (an organization committed to promoting theistic evolution) is not quite clear on the ontological status of miracles. On the one hand, it states that "at BioLogos . . . we accept miracles"; on the other hand, however, they say there is no difference between the supernatural (i.e., miraculous) and the natural action (i.e., through natural secondary causes) performed by God in the universe: "God is just as involved in the regular patterns of the created order as in miracles"; "The [biblical] authors do not make a distinction between natural and supernatural events." This implies that the miraculous events are actually reducible to natural events. See https://biologos.org/common-questions/does-modern-science-make-miracles-impossible. The same type of confusion was promoted by the late Archbishop J. Zycinski, who believed that there was no distinction between natural and supernatural events: "All the world is one great intervention of God." See Heller and Zycinski, *Dylematy ewolucji* [*The Dilemmas of Evolution*], 154n.

universal common ancestry and the natural emergence of species.[9] Hence, theistic evolutionists adopt biological macroevolution as something of a paradigm, rather than a conclusion from empirical studies.

Theistic evolution is sometimes called "evolutionary creation." This is an oxymoron, because evolution belongs to the natural order, whereas creation transcends this order. One and the same thing cannot be produced by a natural secondary process and a direct divine act. Therefore, if something evolved, it could not have been created, and vice versa.

The idea of theistic evolution is depicted in Diagram 17B. The circles represent separate natures or natural species. The thicker line represents the single act of direct causation that initiated the first living being or the universe as such (e.g., the Big Bang). The thinner lines represent natural secondary causation, which generates subsequent natures.

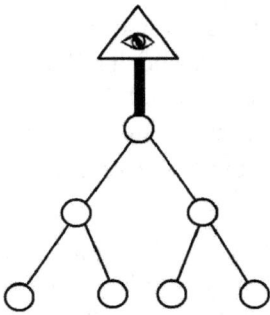

Diagram 17B. Theistic evolution. The bold line represents direct divine causation; the regular lines represent secondary causation.

3. *Progressive creation* (PC) agrees with theistic evolution that the universe was created out of nothing at the beginning of time. It also accepts an old universe (the concept of deep time), spanning billions rather than thousands of years. However, progressive creation states that God

---

9. This thought-pattern was already established by one of the first Catholic evolutionists, St. George Jackson Mivart. In his main work (*On the Genesis of Species*), he dismissed the Darwinian mechanism of evolution by natural selection, yet he was all in favor of universal common ancestry and natural transformation of species. Today it is a still a common belief among theistic as well as atheistic evolutionists that the big Darwinian postulates (such as universal common ancestry) remain valid (they are "facts," as they call them) even if the mechanism of evolution proposed by Darwin does not explain evolution, or is experimentally found insufficient. See, for example, Laurence Moran, "Evolution Is a Fact and a Theory."

created different species (according to our understanding of species in II,I,a) separately by directly exercising his power over vast periods of time. Moreover, it states that creation was definitively completed with the creation of man, but that God nevertheless continues to maintain creation in existence (*conservatio rerum*) and works in the history of salvation. These actions, however, cannot add anything like entirely new natures to the created order.

4. *Young-earth creationism* (YEC) does not differ from progressive creation in its essential understanding of the origin of species. Similarly to theistic evolution and progressive creation, it accepts a first creation at the beginning of time. However, in contrast to theistic evolution and progressive creation, it adopts a short timescale for the universe, spanning thousands rather than billions of years. Hence, the whole history of creation occurred within six days understood as natural days, and the whole history of the universe is not longer than a few thousand years (about six thousand).

Sometimes young-earth creationism is called "scientific creationism." This name, however, is an oxymoron of the same kind as "evolutionary creation." The idea of creation is theological and states that species emerged supernaturally. Since it describes the supernatural actions of God, it cannot be addressed by science, and thus by definition it cannot be scientific. "Scientific creationism" sometimes refers to the endeavor of creationists to dismiss deep time by an appeal to various observations and interpretations derived from science. For this reason, the term "scientific creationism" may be used in a historical sense as referring to those attempts.[10]

Progressive creation and young-earth creationism can be seen as variants on the basic notion of "special creation." The idea of special creation is depicted in Diagram 17C. Direct divine causality is represented by the thicker lines, and secondary (instrumental) causality by the thinner lines. The diagram shows the separate creation of species, which propagate according to their kinds owing to secondary causes, such as natural generation. We see that neither progressive creationism nor young-earth creationism excludes or diminishes secondary causation. Rather, they restore the meaning and significance of primary causation.

---

10. Scientific creationism of this kind may be linked to the book by Whitcomb and Morris, *The Genesis Flood.*.

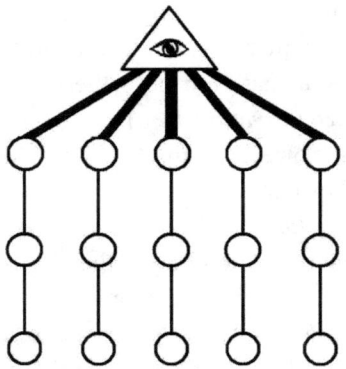

**Diagram 17C. Special creation. The first representatives of each species or kind are created directly by God.**

The fact that the core idea of special creation, i.e., direct creation of species, is not scientific does not mean that it is irrational or antiscientific. If direct divine causation actually accounts for the origin of species, then science should be unable to demonstrate the natural emergence of species, and thus scientists should be open to a theological interpretation. Creationism can be scientific only in a negative sense, by showing that there is no evidence for the evolutionary origin of species. Positive arguments for creation can come only from theology. For this reason, special creation (whether progressive or young-earth) belongs to the domain of theology and should not be called scientific.

### c. The Inaccuracy of the Term "Creationism"

Creation is a theological idea whereby some elements of the universe (particularly different species of animals and plants) were formed supernaturally and directly by God. Usually the suffix "-ism" is added to concepts that try to explain the whole of reality with reference to one well-defined principle. For this reason, "-isms" usually end up as some kind of broader reductionist philosophy. The belief in creation, at least in its Catholic understanding, does not seek any principle that would explain all reality. It is just one (and probably not the most important) element of the whole set of beliefs that constitute the Catholic faith.

Moreover, the theological concept of progressive creation accepts all scientific data and well-demonstrated theories. For this reason, belief in progressive creation should not be called creationism, just as the belief in the Resurrection is not *resurrectionism* or the belief in the virginal conception of Jesus is not *virginism*. Belief in creation is not an ideology that seeks to conquer human minds. It is simply one article of the Christian creed that today has gained even more credibility and support in natural knowledge than ever before.

In contrast, young-earth creationism challenges the overwhelming scientific evidence supporting deep time. In this sense, young-earth creationism is antiscientific. Creation in young-earth creationism becomes a universal principle that attempts to explain all of reality. Since it reduces science to the Bible (specifically, it replaces scientific evidence with the biblical text), it falls into the category of reductionism. And for this reason it can be rightly called creation-*ism*. We propose, therefore, to distinguish creationism from belief in creation. The first refers to young-earth creationism, while the second to an article of Christian faith consisting of the direct supernatural formation of species and man. In the latter sense we speak about *progressive creation* rather than *progressive creation-ism*.

### d. The Interpretative Problems with Theistic Evolution

We suggested (in III,1,b) that theistic evolution in its most popular form is prone to pantheism. Indeed, it is not easy to define it without falling into non-Christian ideas such as deism, pantheism, or emanationism. One interpretation may assume that guided evolution is the work of both God and natural causes, which complement each other so perfectly that the supernatural influence is not noticeable in the scientific realm. In this sense, evolution would be a long series of minute miracles that create the impression of a physically continuous event. The leaps in nature created by these miracles are so tiny that they cannot be detected by the scientific method. An example of this approach comes from those theistic evolutionists who say that God works in nature at the sub-atomic level.[11] The

---

11. A classic example of this approach is found in the works of Robert Russell, whose ideas are summarized in his BioLogos article, "Miracles and Science: A Third Way." According to Russell, his "QM-NIODA" (quantum-mechanical, non-interventionist objective divine action) is not interventionist and occasionalist, because science cannot detect divine action and from the scientific point of view all natural phenomena are as continuous as they would be if God did not work on the subatomic level.

quantum level for science is the realm of chance. Since God works below the lowest level that can be studied via the scientific method, his activity is undetectable. At the same time, these apparently random forces ultimately generate the preplanned effects so that to the human observer, order simply emerges from chaos.

This interpretation, however, assumes the presence of supernatural divine actions in the natural order; therefore, it is not theistic evolution. It boils down to a specific form of occasionalism where God causes each event in nature by using the sub-atomic level of physics. This interpretation has in common with theistic evolution the idea of continuous creation, but it strays from theistic evolution by adopting supernatural divine activity in the process of evolution. Even though this activity is undetectable by science, formally speaking, it violates methodological naturalism—the basic assumption in theistic evolution. Owing to the fact that on this view God works through matter, it falls under the charges that theistic evolutionists typically file against any form of creationism, namely, interventionism, "god of the gaps" and occasionalism.

Some theistic evolutionists believe that God and nature work simultaneously on two levels that parallel each other to the perfect degree. Thus, an evolutionary event like the emergence of a new species is not explicable within nature itself, but it can happen thanks to "divine concurrence." Scientifically we only have access to the level of nature, but theologically we study the level of theology. Thus, one can consistently believe in natural emergence of species (on the scientific level) and their natural emergence studied by science. This approach is close to the idea of NOMA discussed before (I,1,d). But this interpretation is clearly wanting, because we do not know how the two levels are supposed to connect. This interpretation does not explain how divine action would be realized in nature. It actually ends up disconnecting nature from the divine, which is a form of deism—after the initial creation, God does not act upon nature in any realistic sense.

Another interpretation of theistic evolution assumes that both God and nature cause the same effects. For example, an event involving seven coordinated mutations that produce a new protein function is not possible in nature (see II,5,f). Theistic evolutionists would agree that nature alone cannot produce this effect, but nature *guided by God* can. So, the

---

But the fact that something "looks continuous" does not mean that it *is* continuous. On Russell's account, NIODA is a good example of occasionalism even if undetectable by science.

seven coordinated mutations can happen, because they are an effect of the simultaneous work of nature and God—God influences matter in an invisible way so that from a biological perspective it looks like random mutation but theologically it is a work of God.

Apart from a complete lack of empirical evidence for this idea, the problem with this approach is that it melts two causes down to one: if one effect is produced by two causes simultaneously and in the same respect, these two causes must be identical; therefore, they must be one. In other words, if there is one causative action there must be one causative agent. Thomas Aquinas confirms it when he says, "It is impossible for two complete causes to be the causes immediately of one and the same thing" (*Summa Theologiae* [hereafter *S.Th.*], I,52,3,c). Hence, this interpretation of theistic evolution boils down to monism.

Monism has three forms: materialistic, when God is identified with nature; panentheistic, when nature becomes a part of God; and pantheistic, when nature is identified with God. When theistic evolutionists say that God guides or is present in the process of evolution, they fail to adequately explain how exactly God's causation would differ from the sort of causation found in monism. Christian evolutionists do not offer a solution to this problem. In our opinion, such a solution does not exist. Theistic evolution must ultimately fall into either deism (when evolution is initiated by God at the beginning of time) or monism of some kind (when God is present in evolution).

### e. The Amount of Creation in Creation

Thus far we have referred to creation or special creation as the separate production of entirely new forms of living beings performed directly by God. In fact, however, even these authors who reject theistic evolution may greatly differ regarding what must have been created versus what could have emerged naturally. And historically speaking, even theistic evolutionists differed on this point. Even Darwin, after arguing all through *The Origin of Species* that the diversity of living beings was produced naturally, in the conclusion of the book left the vague impression that he thought God had created one or a few initial organisms.[12] The

---

12. However, note that in the *first* edition of *The Origin of Species*, Darwin did not explicitly say that God created the first form or forms of life, but only that "There is grandeur in this view of life, with its several powers, having been originally breathed into a few forms or into one" (490); the additional words "by the Creator" were not

first Catholic evolutionists (St. George Mivart, Raffaello Caverni, Dalmase Leroy) excluded the possibility of the human body being formed by evolution. Most of them also assumed the creation of the first form of life. Later, however, scholars such as John A. Zahm, Henri Dorlodot, and Ernest Messenger speculated over the possibility that evolution produced a body capable of receiving the human soul. They also accepted the spontaneous generation of the first life from inanimate matter.[13] In recent times, a document from the International Theological Commission (ITC) adopted a purely naturalistic explanation of the whole history of the universe, with the exception of the human soul and the beginning of the physical universe (*creatio ex nihilo*).[14] But it is not rare for contemporary theistic evolutionists to support the multiverse hypothesis, which is aimed at destroying the scientific case for the temporal beginning of the universe and creating a new point of conflict between the biblical faith and the claims of scientific community. (In fact, multiverse is not a scientific theory, but it is presented by scientists, which gives it an appearance of science.[15]) Furthermore, there are a few Christian scholars who propose that the human soul is also an effect of evolution, namely, evolution of the brain. Thus, contemporary theistic evolution is not in balance

---

inserted until the second edition (also on 490), though they were maintained in subsequent editions up the to sixth and final edition (on 429). But even the added explicit reference to the Creator does not mean that Darwin actually supported this view. There may be "grandeur" in many opinions. For instance, one may say that there is grandeur in the view that all men are wealthy and healthy, but it does not follow that the statement is true. And just two pages earlier, Darwin had written: "Authors of the highest eminence seem to be fully satisfied with the view that each species has been independently created. To my mind it accords better with what we know of the laws impressed on matter by the Creator, that the production and extinction of the past and present inhabitants of the world should have been due to secondary cause" (488, first edition). So on the question whether the first life forms had to be specially created, or emerged naturally from "laws impressed on matter," Darwin does not provide a clear and consistent view. On a strict reading of the *Origin of Species* one may consider Darwin a pure atheist and nothing of what he says would contradict this interpretation. For comparison of the text of differing editions of Darwin's works, see the Darwin Online website at http://darwin-online.org.uk/contents.html#origin.

13. For more details, see Kemp, "God, Evolution, and the Body of Adam."

14. See *Communion and Stewardship*, no. 63. For more about the history of the first Catholic evolutionists, and for our position on *Communion and Stewardship*, see *Catholicism and Evolution*, 85–136 and 244–52, respectively.

15. Multiverse theory implies that the Big Bang doesn't have to be the only beginning of the universe. It may be just one of multitudes of great explosions and collapses that happen occasionally in one or more of the universes throughout eternity.

regarding the "amount" of naturalistic explanation for origins. The most common boundaries of naturalism are those adopted by the ITC—initial creation out of nothing and then special creation of the human soul. In this way, theistic evolution removes God's supernatural causality from the visible realm and places it entirely in the invisible.

An analogous problem involves scholars who generally reject theistic evolution and believe in creation instead. But even they encounter problems with defining "how much creation there is in creation." As we pointed out above (II,1,a), since the time of Linnaeus species fixism has been almost entirely abandoned. Today, only extreme young-earth creationists would claim that biological species or even varieties were created in the exact forms they have today. Over the past century, different scholars adopted different limits of evolution. For example, in the nineteenth century an American naturalist Louis Agassiz maintained that biological species and even different human races have separate origins. At the other end of the spectrum, an Austrian scientist and clergyman, Erich Wassmann, SJ, developed the idea of polyphyletic evolution—the concept that all living beings descended from a few dozen originally created types. Polish philosopher Mieczyslaw A. Krapiec claimed that supernatural creative acts must have taken place *at least* in the transitions between inanimate and animate matter, then between vegetative and sensory life, and then between sensory and intellectual life. A similar solution was proposed by Thomists, such as Charles De Koninck and Jacques Maritain.[16] This, however, places both scholars among the theistic evolutionists, because their concept implies the natural origin of species. As we can see, different authors have different opinions regarding what must have been created and what could have emerged naturally.

The proper understanding of creation should satisfy all of the requirements and challenges posed by all three levels of human knowledge. For example, if we assume only the creation of matter and the human soul, we cannot read Genesis literally and historically, i.e., in accordance with the Catholic tradition. Hence, even if this solution were compatible with science and philosophy (which is not the case), it would still lack theological support. If we assume the creation of just four substances (material, vegetative, sensory, and spiritual), we still encounter serious problems from theology and philosophy. This idea may overcome the problem of

---

16. Krapiec, *Wprowadzenie do filozofii* [*Introduction to Philosophy*], 256–65; De Koninck, "The Cosmos. The Philosophic Point of View," 256–321, 258; Maritain, "Toward A Thomist Idea of Evolution," 85–131.

the sufficient cause, but it fails to resolve all the other philosophical problems described in the previous chapter (see II,3). In particular, we still do not know how the accumulation of accidental changes could generate entirely new substantial forms. This idea is also incompatible with the book of Genesis, which teaches about creation of animals and plants according to their kinds (which implies that there are more than just three kinds of living beings). Moreover, all of these solutions are incompatible with biology. One of the many reasons for this is that genetic information in different taxonomical families contains a number of unique genes for each family or even biological species. And as we demonstrated above (II,5,b), even a moderately complex gene cannot be invented by means of the neo-Darwinian mechanism. Hence, the only solution that meets the requirements of all three domains of knowledge is the separate creation of species, understood as "natural species" (II,1,a).

However, at this point a skeptical reader may ask a legitimate question: Why should the minimum of creative acts be established at the level of natural species—why not move on to include more and more events of natural history? Why should we not accept the extreme position of species fixism, or postulate the direct creation of mountains, river beds, or different breeds of dogs? Why shouldn't we finally end up in occasionalism, saying that God creates every event in every moment by a direct act?

The explanation for why creation should not expand (and thus turn into creation-*ism*) comes from science. Science tells us what nature can accomplish. And if something can be accomplished by the ordinary operation of nature, it should not be attributed to creation. This is why Thomas Aquinas says that "in the works of nature creation does not enter, but is presupposed to the work of nature" (*S.Th.* I,45,8,c). For example, God did not need to create fully grown trees, because if we plant a seed, assuming that water, soil, and light are present, the seed will grow naturally, without any supernatural assistance on the part of God. However, the first such seed cannot naturally emerge, because it requires a coherent set of different types of information, including a significant quantity of unique genetic information that cannot be produced by any known physical law. The tree is a separate, distinguished nature. As another example, dog breeds can be attained by guided selection; thus, we should not assume their special creation. In contrast, most fertilized eggs from which animals originate require a parent to carry the egg and thus allow for the offspring to be born alive. In such cases a species cannot start with a seed but needs to be originated with a pair of adult specimens who are able to propagate under natural conditions. This is why the scholastics believed that the so-called "higher animals" must have been created in the adult form, whereas the "lower animals" and plants could have been created in the form of seeds.

The list of examples could be continued, but this is enough to present the principle of how to recognize the limits of creation without falling into scientifically untenable extreme fixism or theologically untenable occasionalism. Once we take this principle into account, belief in special creation is not as ridiculous or incredible as it is commonly considered in our times. A summary of our discussion about the limits of creation is presented in Diagram 18.

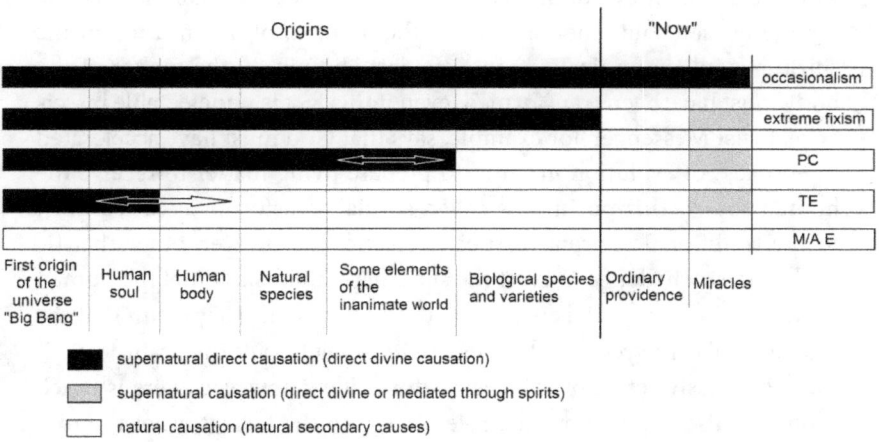

Diagram 18. Different "amounts" of creation in natural history according to different positions.

In Diagram 18, we see a comparison of five concepts that differ in terms of the scope of divine supernatural activity in two domains—the past (the history of creation) and the present (the time of providence). The past is referred to as "origins" because the events considered in this category respond to the question *from where?* rather than *how?* (see I,6,a).

The diagram addresses three types of causality. The first is supernatural and direct, which means that God does something directly without the active help of any creature, either spiritual or material. The second is supernatural causation in a more general sense. This means that God works in physical reality in a way that transcends the power of nature. This action may be either direct (as in the first case) or mediated by supernatural beings (namely, angels or souls). The third type of divine causality is the general providence in which God acts as the final cause of all creation and the One who maintains the universe in existence. This type of causality does not produce supernatural effects in creation. In the third type of causality, the actions of nature are not surpassed, suspended, or overridden by supernatural power. We see that progressive creation (PC) is a middle position between two extremes—occasionalism, which completely excludes secondary causation, and materialistic/atheistic evolution (M/A E), which makes no room for any supernatural causality.

The arrows pointing both directions indicate possible variants of the presented positions. Thus, some proponents of theistic evolution (TE) would claim that evolution cannot produce a body suitable for receiving the human soul. They say that in the moment of the infusion of the spiritual soul (hominization) into the last "hominid," the matter of the body must have been transformed, even if this fact is undetectable by science. Ernst Messenger, for example, says that God must have accelerated the process of evolution in order to prepare living matter to receive the human soul.[17] This position is called special transformism (in contrast to the traditional concept of special creation). However, other theistic evolutionists challenge not just the supernatural formation of the human body but also the special creation of the human soul. They claim that the human soul (or psyche) is the effect of the evolution of the brain.

Progressive creation is a well-defined idea; however, there is much room for discussion about the role of God in the formation of the major features of nonliving physical reality (the first planetary systems, the earth, the first geological formations, and so on). These may be attributed to natural causation or supernatural causation, or direct divine causation. Since this problem is not relevant to our discussion, we do not address it here (see also I,6,b).

### f. Who Debates Whom in this Debate?

We see that the four positions in the debate over the origin of species (M/A E, TE, PC, YEC) stretch from the purely naturalistic concept of materialistic evolution at one extreme to the largely fideistic idea of young-earth creationism at the other. The materialistic position refers only to scientific data, thus excluding theological knowledge ($t_2$ in Diagram 1),

---

17. Messenger, *Evolution and Theology*, 93.

whereas the fideistic position derives facts about nature primarily from the Bible, sometimes seeking connection with science in the form of so-called scientific creationism.

The two middle positions take into account both science and theology; however, theistic evolution gives primacy to science, while progressive creation gives primacy to theology. The perspective of these two is different: Theistic evolution derives its knowledge from the scientific theory (not evidence) and then looks for justification and connection with religion; the main question for theistic evolution is how to reinterpret the biblical message and Christian tradition in order to make them compatible with the story told by the majority of biologists. In contrast, progressive creation derives knowledge about origins from the Bible and then critically scrutinizes theories presented in science in the light of faith (and scientific evidence).

Each of the four concepts addresses two separate questions: the first one concerns the *timescale* of natural history, and the second one concerns *how* species emerged. Regarding the first question, all theories except young-earth creationism adopt the overwhelming scientific evidence supporting a universe whose history stretches back for billions rather than thousands of years. Regarding the second question, atheistic and theistic evolution accept the evolutionary (i.e., natural) emergence of species, whereas the creationist theories speak about their supernatural formation by God. These are the basic similarities and differences among the four concepts of origins.

We should also notice that debates over the origin of species fall into two types. The first applies to mainstream culture. It usually involves the division between materialistic (or even more commonly, atheistic) evolutionists and theistic evolutionists. The main line of argumentation in this debate consists of atheists trying to prove how chance can do anything and theists trying to superimpose a little bit of religion on an essentially naturalistic paradigm. They do it by claiming that God can use chance events to obtain His preplanned effects. Because theistic evolutionists normally do not acknowledge the possibility of recognizing design in nature by means of the scientific method, their arguments fall on deaf ears. Atheists and materialists who oppose theistic evolutionists simply do not care about their philosophical and theological arguments, because they do not believe in anything but matter. Consequently, the theistic evolutionists' narratives about teleology perceived on a metaphysical level of nature or God's invisible guidance of chaos sound like pious fairy tales to staunch atheists. Theistic evolutionists make attempts to break the atheists' naturalism, but this is not possible because theistic evolutionists themselves adopt the same kind of naturalism, albeit to a lesser degree (first-grade naturalism according to our typology in I,2).

The second debate takes place within the Church. The two parties of this debate are usually theistic evolutionists and young-earth creationists. The debate primarily concerns scientific issues. Theistic evolutionists try to convince creationists to accept the same "scientific evidence" that they have adopted from the atheistic evolutionists. Creationists respond with a thorough critique of the alleged evidence supporting evolution. But they immediately add their own "scientific truth" about the very short existence of the universe. In this way they become an easy target for both theistic and atheistic evolutionists. Evolutionists do not need to respond to the scientific critique of biological macroevolution. Instead, they simply focus on ridiculing creationists' claims about the young earth. In this way, the defense of creation offered by the young-earth creationists remains futile. Ironically, young-earth creationists, by rejecting the evidence for "deep time," end up in the same kind of rejection of facts that makes evolutionists believe in probabilistic miracles (see II,5,a). The saying that opposite extremes end up in the same errors is apparently true in this case.

As we can see, progressive creation is typically overlooked in both secular and ecclesiastical debates. Indeed, theistic and atheistic evolutionists, as if by default, do not distinguish between progressive creation and young-earth creationism. Sometimes this inability to distinguish between these two seems to be part of the evolutionists' strategy for evading a confrontation with the scientific critique of their position. Evolutionists push all proponents of creation into one corner labeled "creationism," and then they show how ridiculous is the belief in the young universe, instead of explaining how the evolutionary mechanism is supposed to account for the origin of species. Since evolutionists believe that biological macroevolution is a "scientific truth" or a "fact," they consider any contrary idea antiscientific, or simply irrational. This is why, for evolutionists, all critics of evolution are "young-earthers," and these in turn are regarded as equivalent to "geocentrists" or "flat-earthers." By conflating pseudoscience and reasonable scientific critique of evolution (such as is found in intelligent design theory), evolutionists seek to ridicule everyone who does not accept their naturalistic position. In this way, they excuse themselves from the academic debate that otherwise would include a much harder task of defending the alleged evidence for biological macroevolution.

The ecclesiastical debate over all three theistic positions (TE, PC, YEC) awaits opening. It should proceed in such a way that the positions are not misrepresented or confused with one another. This is both a moral requirement for any honest discussion and a formal requirement for any truly academic debate. Only under these conditions will the debate bear fruit and lead people of good will closer to the truth about origins.

## 2. Christianity and Creation

### a. Two Christian Interpretations of Genesis

Having presented the current positions in the creation-evolution debate, we will now focus on the past and try to recover the traditional understanding of creation as taught by Catholic theologians before Darwin. It should not surprise anybody that the Church used to have a well-defined teaching on origins. However, the content of that teaching may be surprising. The reason why many contemporary scholars consider the original Christian doctrine implausible, strange, or even ridiculous is that it has not been taught, defended, or explained for almost a century now. In the Church, tradition (including scholarly and oral tradition) plays a significant role. If one generation of theologians ceases to know some part of tradition, there is nobody to pass this knowledge on to the next generations. Today, we have probably the fourth or fifth generation of theology professors, priests, and catechists who simply do not know the classic dogmatic tradition expressed in the phrase *de Deo creante et elevante*.[18] The following subsections (III,2,a–c) are designed to present a brief account of the traditional Church position on creation.

---

18. This phrase refers not simply to a single treatise, but to a whole theological subdiscipline explaining God's creation of the universe and the human race. Today this area is more often called protology (the "science of the beginnings"). Throughout the theological era of treatises (tracts) and manuals (from the sixteenth through the early twentieth centuries), there were many authors writing books entitled *De Deo creante* (or *De Deo creante et elevante*). Among the most recognizable were the representatives of the Roman neo-scholastic and neo-Thomistic schools, e.g., Joseph Pohle, Charles Boyer, Dominic Palmieri, Camillo Mazzella, Christian Pesch, Giovanni Perrone, and Adolphe Tanquerey.

When St. Thomas Aquinas summarized the Christian understanding of Genesis, he spoke about two interpretative traditions, one dating back to St. Ambrose and the other to St. Augustine. Both traditions assume that Genesis is a book that recounts the true history of creation, even if it does not do it in a detailed or methodical manner. According to the first tradition, there was a succession of time in creation. According to the second tradition, all things were created in the beginning, in one act, directly by God, but some things in their fully developed forms and others in the form of seminal reasons (Latin *rationes seminales*, Greek *logoi spermatikoi*). These "seeds" developed their full natures only later, after creation was completed. According to Augustine, Moses divided one divine act of creation into six days because the simple-minded recipients of his text would not comprehend the actual simultaneous creation of all things.

Aquinas affirms that the Ambrosian tradition of progressive creation is more common in the Church (i.e., held by the majority of theologians), and more congruent with the Bible (at least according to a superficial reading of the text), but also more vulnerable to the critique of unbelievers. The Augustinian tradition, in turn, is less common, less compatible with the surface sense of the book of Genesis, but more resistant to the attacks of unbelievers. Aquinas even says that he prefers the Augustinian tradition over the Ambrosian one, but will defend both.[19] The two traditions are compared in Diagram 19.

---

19. Aquinas, *Scriptum super Sententiis, liber II*, dist.12,q.1,a.2c. A thorough description of Aquinas's position regarding the two Christian interpretative traditions may be found in our paper "Thomas Aquinas on Creation, and the Argument for Theistic Evolution," as well as our book *Aquinas and Evolution*, 96–124.

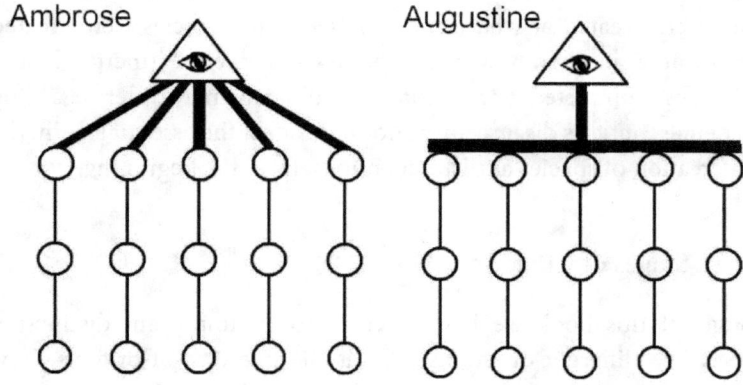

Diagram 19. A comparison between the two Christian interpretative traditions of Genesis.

In Diagram 19, direct supernatural causality is represented by the thicker lines, and secondary, natural causality by the thinner lines. The diagram does not include a time line, God's providence, or the divine maintenance of being. We see that according to the Ambrosian tradition there are different direct acts of God in bringing about different species. These events took place over a period of time that the Bible describes as six days. After the separate species (the biblical kinds) started to exist, they pass on their natures through natural secondary causation (generation). In the Augustinian tradition, all species are created simultaneously, in one direct divine act, at the beginning of time. Later, though, species propagate and reveal their natures through natural generation. Thomas Aquinas says that these two traditions do not differ in what is essential to faith, but they differ regarding accidental issues.

One essential thing is, according to Aquinas, how things were brought about, namely, through direct causation. And this is common to both traditions. The accidental truths are related to (1) the time relationship between the events (whether they all happened at once or over the six days), (2) the understanding of some terms used in the first chapter of Genesis (such as firmament, light, earth), and (3) many historical details (*multa historalia*). For example, it is irrelevant for faith where the Israelites crossed the Red Sea, how many people gathered on the Mount of Beatitudes, how many blind men were restored to sight by Jesus, or which route St. Paul took when traveling to Damascus. Analogously, historical details that are accidental to belief in creation may include how many species were created at what time, whether some species were created before or after another, how many species underwent extinction before creation was completed, and so forth. Theologians may differ regarding these details, but this disagreement does not affect the essential truth: the direct creation of species as distinct entities, from the beginning.

## b. Three Stages of Creation

Throughout this book we have referred to an important distinction between two different questions. One is the question of origins—how things started; and the other is the question of their current operation, *how they are* after creation is completed. Now it is time to explain this distinction in greater detail.

The original Catholic teaching on creation speaks about three stages in the history of the universe. The first stage includes *creatio ex nihilo*—the first act of creating being out of nothing. It is said that in the beginning God created both spiritual and material beings without any precedent matter or potency. After its inception, the material world must have had some form, because without any form, matter cannot exist. However, this form was unspecified. Genesis describes it with the Hebrew phrase *tohu vabohu*, which can be translated as a lack of form and shape, or chaos and emptiness.

Matter, as it is the least actual of all being, cannot generate anything by itself. Therefore, God continues his activity in the second stage, called the *work of formation* (*opus formationis*). In this stage, the scholastics distinguished two phases: the work of distinction (*opus distinctionis*) and the work of decoration or adornment (*opus ornatus*). The work of distinction includes the production of things that are—in a way—one, in the sense that the newly created being is not completely detached or distinguished from the primitive matter. This includes the formation of the seas and the earth (out of the original chaotic blend of water and earth), and also the creation of plants (things which, though distinct, remain attached to the earth). The second phase—the work of adornment—includes the creation of things that are detached from others and can move independently, such as the lights in the heavens (sun and moon and stars), animals of air, water, and land, and man. After man is created, the work of creation is finished once and for all. Even the new creation foreshadowed after this world ends will have a different character; it will be a re-creation rather than creation. No new natures (species) can appear after God finishes his creative activity on the sixth day.

The third stage of the universe refers to the history of salvation, which begins with the fall of man. This is the time in which we live. Even though God does not create anything new, he still acts supernaturally (directly or indirectly) in the universe through miracles, special revelations, prophecies, visions, infusions of grace into human souls, etc.

There is also one more type of divine action that underlies the two fundamental eras of the universe, that is, the history of creation and the history of salvation. This divine action is God's maintenance of the universe in existence (*conservatio rerum*) and His ordinary providence, which leads all natural things to their natural ends. Whatever begins to exist—whether the universe as a whole, in the first act of creation (*creatio ex nihilo*), or any particular nature (such as different species)—immediately falls under this type of divine activity, or otherwise would completely cease to exist. Diagram 20 summarizes the different types of divine causality in time.

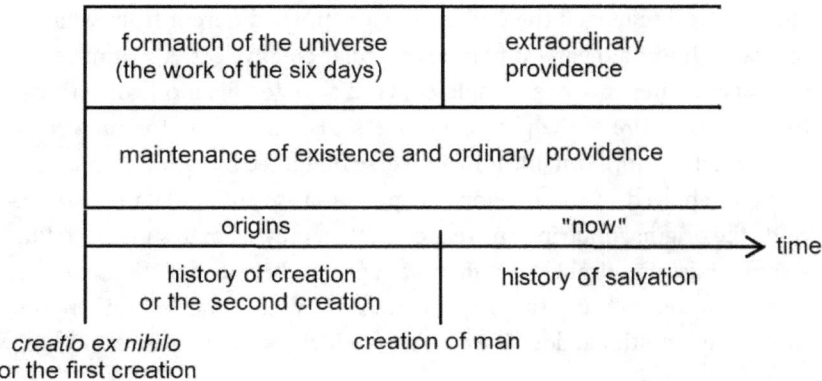

Diagram 20. Different types of divine causality over time.

Now, whenever previously we referred to the distinction between "origins" (or "the past") and "now" (or "our times"), we had in mind this primarily theological distinction between the order of the formation of the universe and the order of the subsequent operation of the universe under ordinary providence. The creation of Adam is the last work of formation (strictly speaking, adornment), because man is the last specified and distinct nature to be created. The creation of Eve maybe interpreted either as the last act of creation or as the first miracle, that is, an act of supernatural providential care. In either case, the creation of the first woman has a very special status in the whole order of creation.

As we said, this classical Catholic understanding of creation may seem incredible for contemporary believers. But this is not because the doctrine itself is strange, irrational or implausible. The reason why it is vulnerable to rejection is that today's Christians exist in a culture pervaded by naturalism. If we consider this doctrine with an open mind, we will realize that the traditional belief makes perfect sense.

Before man appeared on earth, the universe had no rational inhabitants who could observe it. Over billions of years, everything was being prepared for man, who was not yet present. We tend to judge everything from our own perspective, according to how we see things today. And it is true that God's creative activity today would be somewhat incompatible with our existence. We would not like to see new creatures appearing all of a sudden in different parts of the world. The universe would be unpredictable, unstable, not a good place to live in. But before man was created, God managed the universe in a manner different from what we see today. It was a theater for His spectacular creative activity, admired by the angels. After this was completed, God changed his mode of action—from *creative care* of the physical universe, He moved on to *providential care*, which is more suitable for a universe inhabited by rational creatures.

We showed in the previous chapters that scientific data do not exclude the original doctrine on creation; rather, they seem to support it. Philosophy raises a number of difficulties against theistic evolution, and the Genesis account clearly supports the traditional doctrine. Why, therefore, should the traditional doctrine be less convincing than theistic evolution?

## c. The Nature of the Creative Act

In our times, popular and academic discussions abound in errors regarding the understanding of divine creation. Most of these errors are simply a consequence of tinkering with the notion of creation in order to make it compatible with evolutionary cosmology.[20] In this subsec-

---

20. The origin of this attitude may be found in the work of St. George Mivart, an

tion, we will clarify and organize a few concepts regarding the Christian understanding of creation.

One of the common errors consists of saying that God used evolution as a secondary cause of creation. According to this perspective, the evolutionary process is like a tool that a craftsman would utilize to accomplish his work. In response to this, all Christian tradition firmly claims that only God can create (in the proper sense of the word). The act of creation cannot be performed by any creature, either celestial or material, even acting only as an instrumental cause.[21]

To understand why, we need to realize that in every creative act something is produced out of nothing. There is no proportionality between something and nothing. The ontological distance between nonexistence and existence is infinite. And only an infinite being can overcome this infinite ontological chasm. But we must not think of this as going from one state to another. As the scholastics used to say, *creatio non est mutatio, sed simplex emanatio entis ex nihilo*—creation is not a mutation (change), but a simple emanation of being out of nothing. Creation is not a change, because change affects being only in a certain aspect, whereas creation is *productio rei secundum totam suam substantiam*—the production of a thing according to its entire substance. Therefore, creation can be described as an instant emanation of a being without any intermediary phases or any substrate.

Since any created being causes something only in a certain aspect, whereas creation is calling into existence the totality of a being, it follows that no creation can create by its own power. Furthermore, the power of creating cannot be passed on to creation. Again, to understand this, we

---

early Catholic evolutionist. In his main book, *On the Genesis of Species*, he distinguishes creation in its original meaning from creation in a derivative meaning, which refers to the action of nature. Mivart shows how evolution may be reconciled with creation in the derivative meaning. See Mivart, *On the Genesis of Species*, 269. This solution rests upon the equivocality deliberately introduced by the author. Today's theories reconciling evolution and creation usually have roots in the same error.

21. Secondary causes can be divided into two categories. In the first, the secondary cause is moved by the first cause yet works by the power of its own nature. This happens when, for example, a king orders the construction of a palace. The king works as a first cause, an architect as the secondary cause. However, an architect designs the castle by himself, without constant action from the king (the first cause). The second category consists of the so-called instrumental causes. These also are secondary causes, but they cannot complete their task by their own power. For instance, a chisel makes a sculpture, but a chisel is unable to make a sculpture by itself; all its power to make a sculpture comes from the first cause, namely, the sculptor.

need to realize that every creature works according to some principle of operation congruent with its own nature. Man needs a mind, and hands, or tools, or space, etc., to perform any action. Angels need a mind, or will, or at least their own form to perform any action. Any created being is somehow defined and limited in its operation to some means or ways of doing things. But an act of creation presupposes nonexistence; therefore, there is no principle whatsoever whereby a creature could work toward accomplishing this act.

The analogy between God creating and a sculptor using a chisel does not apply to creation because the chisel is of such nature to be used to carve stone; it is its principle of operation to do this kind of work. But no creature is of such nature to create the very being of a thing. If God were to use any tool, any secondary cause, to perform an act of creation, this would be like a sculptor using a plastic knife to cut rock or an engineer changing a house into a car by painting it. God can do it by exceeding any limitations of creatures, but if God does it by completely transcending creaturely limitation, this act cannot be attributed to creatures in any sense; therefore, creaturely action is not a secondary cause of creation. And for this reason, Catholic tradition cannot accommodate the idea that creatures co-create or act in any active way in creation.

Now, the creative act in its most proper meaning is the first creation of all being out of nothing (*creatio ex nihilo*). At that moment, everything was created at once, though without any distinctive forms. These were educed only later, in the work of formation. The subsequent creative actions brought about some novelty that did not exist before. This is especially clear in the creation of plants and animals. Even though the matter for producing them is already present in nature, there is still creation of new substances, new species. God produces these new forms in matter and thus makes entirely new substances (new beings). Hence, in the production of species there is no change or transition, but simple emanation of a new essence out of nothing. And this is why no creation can be a secondary cause in the production of living beings—they could only have been made by God, through his direct power.

Does this mean that species just popped up like rabbits from a magician's hat? This is another great misunderstanding of the authentic Christian doctrine on creation. Genesis 2:19 says, "the Lord God formed out of the ground all the wild animals and all the birds of the air." Genesis 2:7 tells us that also man was molded out of the ground. This means that all the so-called "higher animals" were created in the same manner as man—by supernatural formation from the ground combined with the infusion of the appropriate souls—rational for man and animal for animals (in philosophy these are called immaterial principles or forms, and in the Bible an "element of life," Hebrew *nephesh*).

This mode of creation of species shows that in the second creation (*secunda creatio*, the formation of the universe), created elements may play a *passive role*. The material that God uses to form the universe may be considered a passive cause of the creative action that belongs solely to God. In this sense, God uses creation to bring about new natures as when he uses clay to form the human body. But this is very different from the idea of God passing on the creative power to creatures, or making them participants of the creative act. Also, some scholars, such as Aquinas, tell us that the dust used by God to form the human body was gathered in one place by the angels in the same way that they will gather the ashes of the saints on the day of the resurrection when God will rebuild their bodies.[22] In this case, the angels participate in creation, but again, not in the very creative act, but as helpers preparing matter for the infusion of the totally new form. In contrast, in the first creation (*prima creatio*) not even passive help could have been possible, because nothing but God existed before the first creation.

We said above (III,1,e) that the amount of creation in the natural history of the universe corresponds to the amount of novelty that nature could not have produced by itself. In other words, whatever exists that cannot be produced by nature must have been created. Since the higher animals cannot exist independently as embryos or fertilized eggs, they must have come into existence as adults molded by God in fully grown forms. The situation is different, however, with plants, which can develop from seeds subjected to the basic environmental conditions provided by nature. And this is why, instead of creating fully grown plants, God ordered the earth to bring forth vegetation (Gen 1:11), meaning that God produced the first seeds in the ground.

A quick reading of the biblical text may give the impression that the earth and the waters were asked by God to take some active role in the creation of plants and lower animals. But this is not what the sacred text conveys. To see why Genesis employs this language, we need to take into account at least three things.

---

22. *S.Th.* I,91,2, ad1.

First, the first account of creation (Genesis 1) adopts an objective perspective, as if the events were observed from an external or absolute vantage point. And surely, an external observer could not see the creation of seeds in soil or water. Hence, it was the soil and the water that seemed to bring forth vegetation on land and certain aquatic animals in bodies of water.

Second, once the seed is in the soil, it is left to various environmental factors, such as moisture and sunlight, that trigger development. Thus, it is literally true that the earth "brings forth" vegetation, though it requires seeds in order to initiate this process.

Third, God used ground in the formation of the animals, and thus it is possible—and perhaps even more fitting—that God would use ground to form seeds in the soil, as well as water to form seeds in the water. In any event, the point here is that neither earth nor water played an active role in the emergence of the seeds. The activity of earth and water began only after new distinct substances, in the seminal form, appeared in them.

In order to further specify the nature of the creative act, we need to distinguish it from a few other types of divine work. The first act of bringing the physical universe into existence constitutes *creation* in its most proper sense. The subsequent divine acts belong to supernatural and direct *formation*. More specifically, *creation* refers to the production of all being out of nothing, and *formation* to the supernatural infusion of specified forms into matter whereby new natures are founded. After species are formed, they propagate through natural *generation*. Generation therefore, is distinct from creation, since the latter is an act that transcends the ordinary works of nature and multiplies species, whereas the former is an ordinary work of nature by which individuals multiply within their species. Furthermore, the creation and formation of the universe is not the same as God's *providence* (*gubernatio rerum*). The former brings about new natures, while the latter only directs natures to their ends (i.e., to fulfillment). In creation, the dignity of beings is manifested by the fact that God works directly upon the physical universe, touching it, as it were, with his hands. The providential care also reveals the dignity of beings, because God uses them as secondary causes in the realization of his plans.

The distinction between the divine act of creation and divine providence is neatly grasped by Thomas Aquinas. The Angelic Doctor explains that there are two perfections to every being—one which takes place in creation and perfects the being substantially (i.e., according to its nature), and the second which occurs after creation is completed, and makes the being achieve its proper end.[23] The origin of species belongs to the first type of perfection and thus cannot be the work of providence; the latter however, directs all natures toward their final end, which for the human species is salvation in heaven.

Another crucial distinction separates creation from the *conservation* of things. Theistic evolutionists often allude to what they call *creatio continua* (continual creation), which means that creation is not just a one-time event, but rather a constant action accounting for both the evolutionary origin of the universe and its continual existence. The idea of continual creation does not appear in Aquinas's writings, nor in classical Christian doctrine on creation. Instead, Christianity distinguishes between *creatio* and *conservatio rerum*. The first is the direct and supernatural act of God that brings about some novelty—either the whole universe without the distinction of particular forms (*prima creatio*) or new, specified forms (*secunda creatio*). After things are created, they are maintained in existence by a constant divine operation called the conservation of things. This is an immediate continual action of God that differs from both providential care and creation. By this operation God does not create anything new.[24] From the standpoint of metaphysics, divine conservation of things keeps substance and existence together, but it does not affect substantial or accidental forms in any way. Aquinas compares divine conservation of things to light that pervades air—God infuses being into all things (all substances) as long as they exist; however, this action does not affect the changes (the motion) that constantly occur in the universe.

This is even clearer if we take into account the modern scientific evidence showing that the entire natural order disperses over time in ever-growing entropy, leading eventually to the cessation of all physical activity and form. The fact that all things tend to "perfect" chaos proves that conservation of material being is nothing more than infusion of being into substances without affecting the forms. If the opposite were true, that is, if there were anything like continual creation that produces ever higher orders of nature and new species (as theistic evolutionists imagine) then composite beings would not be susceptible to degeneration and

---

23. *S. Th.* I,73,1,c.
24. Cf. *S. Th.* I,104,1 and *Summa contra Gentiles* III,65.

species would multiply rather than die out, which is not the case in the universe we know.

Finally, divine creative acts were not *miracles*. In a miracle, God suspends or transcends laws of nature in order to bring about his intended end. But the end of a miracle is never a production of a new nature. In miracles, nothing new begins to exist—rather, the created order is restored or perfected, such as restoring sight to the blind or life to the dead. The same applies to miracles of multiplication: bread and fish were multiplied by Jesus, but there was no creation of new species. So, even in miracles of multiplication, the number of natures does not increase; rather, the number of individuals of a given nature is multiplied. Miracles and creation also differ in that miracles belong to the supernatural, but not necessarily to *direct* causation. Unlike creation, miracles can engage as secondary causes angels, the souls of the saints, living people, animals, or even non-living matter.

The distinction between creation and miracles helps us avoid another error common in today's debates over evolution. Many scholars, including many creationists, believe that creation is some kind of *intervention* of divine power in the natural order. Often creationists advocate what they call "interventionism"—the idea that in creation God disrupts or ruins the natural order by his overwhelming power. In response, theistic evolutionists combat interventionism along with the idea of God acting externally on the created order. As they say, "God is not a magician playing with a magic wand," "God does not intervene," "creation is not a series of miracles," "things do not just pop up into existence," and so forth. Unfortunately, in this exchange, creationists create a caricature of divine action and theistic evolutionists combat a straw man. Neither of the parties adopts the classical Catholic doctrine.

To further explain this problem, we need a broader picture. The word "intervention" comes from Latin *inter-venire*, which means "coming in between." It refers to a chain of causes and effects that is interrupted by an external agent (God) who enters in between natural events in order to change their direction and the final outcome. This kind of divine activity is clearly seen in the order of supernatural providence or the history of salvation, after creation is completed. We see divine interventions in the history of Israel, we see them throughout the New Testament, and we see them in the lives of the saints and the history of the Church. Divine interventions make sense in the lives of rational creatures, who choose their goals by the power of their wills. The whole history of salvation is about God bringing humans to their ultimate end—an end that may not be fully recognized by them while they are on the way.

By entering the cause–effect chains generated by human freedom, God accomplishes several goals. First, he reveals to humans their proper end and how to obtain it. Second, he demonstrates that humans are not absolute masters of all reality; instead, they are dependent on a higher being. Third, God helps humans accomplish things necessary for salvation that may transcend their capabilities (such as we find in the Annunciation). In contrast, the creation and formation of the universe happened before man came into existence, so any intervention in the order of creation would lack any teaching value. Moreover, the primary reason for the existence of material being is to serve humans; therefore, the goal of the supernatural formation of the material universe is accomplished not by interventions in the natural order but by the creation of man. This is why God did not have to intervene in the natural order before man was created.

Another reason why creation is not intervention comes from the fact that intervention presupposes the existence of causes and effects, whereas creation does not presuppose anything (or presupposes nothingness). Creation is the addition of new being without diminishing or ruining that which already exists. When God was adding new beings in the work of formation, the cause–effect chains already in existence remained in operation.

We can liken this to the work of an artist painting a landscape. First, we see the earth, the sun, and the sky. This is already beautiful and complete. Nothing is missing. However, the picture may be enriched by adding new elements, such as an ocean or clouds in the sky. This picture is even more beautiful and complete. The painter may continue his work and add birds, land animals, fish, and man, which completes the picture. Each new layer of paint adds something to the total beauty of the painting, yet without diminishing anything else. Similarly, when God adds new natures, this does not destroy anything already created. For instance, when a new animal is created, it can change the course of nature by hunting another animal or nesting in the previously virginal habitat. However, this change—this intervention in the previously existing order—does not stem from the very act of creation of the animal, but from the operation of the animal after it starts to exist. The creation of a new nature begins a totally new chain of natural causes and effects that consequently needs to cross the chains already in operation. This is how creative events are complementary to the totality of creation. Therefore, belief in separate creation of species is not the same as belief in interventionism. Table 2 summarizes the types of divine causality found in classic Christian theology.

| THE ORDER OF | TYPE OF DIVINE CAUSALITY | DIRECT? | SUPER-NATURAL? |
|---|---|---|---|
| Creation | Creatio ex nihilo or the first creation | + | + |
| | Opus formationis or the second creation | + | + |
| Providence (gubernatio rerum) | Conservation of things | + | + |
| | Miracles | +/- | + |
| | Providence | +/- | +/- |
| | Natural generation (natural secondary causation) | - | - |

Table 2. Different types of divine causality. "+" stands for yes, "-" for no, "+/-" means that both answers are possible. Note that direct divine causation by definition must be supernatural.

## 3. Christianity and Theistic Evolution

### a. Theistic Evolution and the Notion of Creation

Theistic evolutionists typically accept the idea of creation out of nothing in the beginning ("first creation"), but reject direct divine causation in the formation of the universe. This leads to multiple confusions in understanding different types of divine causality. One common claim for justifying theistic evolution is that Christianity rejects interventionism and occasionalism. Based on what we said above, we can see that this charge misses the point, because the classic doctrine on creation is neither interventionist nor occasionalist (see Diagram 18). Another common objection is that Christianity does not understand creation as a series of miracles. And this is true, but not because Christianity rejects direct causality in creation (as theistic evolutionists believe), but because creative causality, while being direct and supernatural, is not a miracle strictly speaking (see III,2,c). Theistic evolutionists, in the positive presentation of their doctrine, usually state that God uses evolution as a natural secondary cause to produce different species. In this, they confuse the order of creation with the order of divine providence (i.e., *opus formationis* with *gubernatio rerum*). Indeed, theistic evolution effectively rejects the order of creation, replacing it with divine providence or divine government over the natural order. We can conclude that theistic evolution has two problems: one is its misunderstanding and misrepresentation of Christian teaching on creation, and the other its de facto rejection of at least one essential part of this teaching, that is, the divine work of formation (*opus formationis*).

### b. Augustine and Theistic Evolution

Since St. George Mivart and John A. Zahm in the nineteenth century, many scholars have claimed that St. Augustine was a forerunner of modern theistic evolution. As early as 1926, a Catholic philosopher, Michael McKeough, stated that Augustine's doctrine "constitutes a satisfactory philosophical basis for evolution, and merits for him the title of Father of Evolution."[25] In our times, the same assertion has been supported by Catholic scholars, including Jozef Zycinski, Michael Heller, Ernan McMullin, William Carroll, and many others.

To respond to this interpretation, first we need to stress that the very word "evolution" means something different for Augustine and for Darwin (see II,1,b). And it was precisely on account of this difference that Darwin initially did not use the word at all. He introduced it into his primary work, *The Origin of Species*, only in the sixth edition, after the understanding of the word had changed. For pre-Darwinian scholars, including Augustine, evolution meant revelation or uncovering of something that had existed in some hidden form. In contrast, for Herbert Spencer, Darwin, and their followers, evolution meant a creative process that could generate all kinds of biological novelties. For Augustine, evolution was like unpacking boxes, whereas for modern evolutionists, it is like producing

---

25. McKeough, "The Meaning of the *Rationes Seminales* in St. Augustine," 109–10. However, the same author says in another place that "Augustine believed that things appeared with the same forms they had in his day and that those forms were constant," 78–79). I do not agree that Augustine believed in extreme species fixism, and furthermore I do not know how the two quoted statements of McKeough can be put forward without contradiction.

new kinds of devices from other devices. This change in the meaning of the word creates confusion regarding the teachings of Augustine.

Second, for Augustine, interpretation of Genesis was a mystery, probably the most difficult piece of Holy Writ. When commenting on the various passages of the creation account, he repeatedly states that he is not sure how to understand them. Many times he offers multiple possible interpretations, only to conclude that one can choose whichever interpretation one wants.[26] It is safe to say that Augustine did not know exactly how to understand the Genesis account of creation. For this reason, we should not assume that he offers one specified interpretation that could be used as an argument in the current, and quite sophisticated, debates regarding biological evolution.

Having said this, we need to address the third issue, namely, the concept of seminal reasons found in Augustine. The Doctor of Grace was influenced by Stoic and Neoplatonic philosophies. When he encountered the Christian understanding of the origin of the universe, he was troubled by the fact that the subsequent formation of the physical order over six days too closely resembled pagan philosophical systems. God—reasoned Augustine—is a perfect being, and as such, he should not form the universe over time as if amending and perfecting his own works. Moreover, the succession of time seemed to be excluded from creation in the Book of Sirach.[27] In Augustine's mind, creation should be an instantaneous act emanating everything—with distinctions among species—immediately out of nothing. Augustine looked for a metaphysical model that would harmonize the Genesis account (which he considered primarily a true historical description of what happened) with his philosophical preconvictions. The Stoics' idea of seminal reasons came to his aid. Augustine concluded that God created everything in one act at the beginning of time, though some things were created in hidden forms, such that they would be revealed only later, in the course of the ordinary operations of nature.

We see how Augustine's mind-set heavily influenced his interpretation: the abstract and speculative attitude of Hellenistic philosophical

---

26. See, for example, Augustine, *Confessions*, Book XII, no. 20.

27. Augustine based his judgment on the Latin text of Sirach 18:1 that reads: *Qui vivit in aeternum creavit omnia simul*. ("He who lives forever created all things simultaneously.") The word *simul* appearing in Vulgate and pre-Vulgate texts is a translation of the Greek *koine*, which would be more properly translated as "in common" or "without exception" thus indicating the inclusiveness of space/volume rather than time. Augustine's interpretation was reinforced by the inaccurate translation.

reasoning embraced and effectively modified the concrete and historical message of the Hebrew text. As we noticed (in III,2,a), Thomas Aquinas concludes his account of Augustine's interpretation by saying that it is not common in Christianity and not completely compatible with the Scriptures (at least at first glance). Today, we need to add that Augustine's interpretation is also not compatible with the scientific evidence, because different species appear in the fossil record over long periods of time (not all at once) and fully developed according to their proper forms (not in hidden or devolved forms). Aquinas seems overly sympathetic to Augustine's interpretation, and the main reason for it is that he does not know natural history. Without evidence from nature, Augustine's interpretation is more appealing to Aquinas, because it looks "neater" and is less susceptible to criticism.

Since it is difficult to reconcile Augustine's interpretation with both the Bible and the scientific evidence, it should be essentially abandoned. Instead, it seems to flourish. This odd situation occurs because many Christian scholars repeatedly quote Augustine to support theistic evolution. Thus, in order to support their view, they adhere to a worse biblical interpretation (compared to that of Ambrose) that additionally is not compatible with modern data.

But a bigger problem is that Augustine's idea of seminal reasons would help theistic evolutionists only if it were compatible with theistic evolution. So, the question is, does Augustine really help to save theistic evolution for Christianity? In our opinion, the answer is no, because his own concept of the origin of species differs substantially from theistic evolution. We can see this when we compare the following two diagrams, one representing Augustine's interpretation of Genesis and the other representing theistic evolution (Diagrams 21A and 21B).

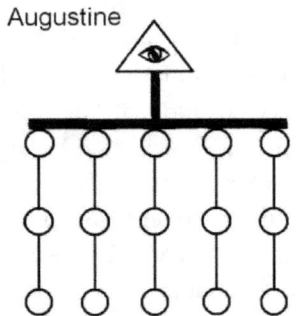 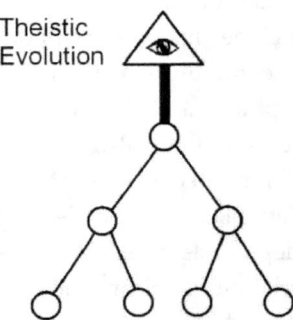

Diagram 21A, Diagram 21B. These diagrams represent the origin of species according to Augustine and theistic evolution. The thicker lines represent direct divine causality, and the thinner lines represent secondary causes, such as generation. The diagrams omit a timescale, God's providence, and the divine maintenance of being.

We see that Augustine's interpretation is incompatible with three ideas that are essential to theistic evolution. First, he maintains that species (at least those of animals) were created immediately by God and not by nature acting as a secondary cause. In contrast, according to theistic evolution, perhaps supernatural divine causation was involved in the creation of the first life form, but usually even this is attributed to natural causes (usually theistic evolutionists believe in the so-called "self-organization" of matter and abiogenesis). Hence, according to theistic evolution, species were formed naturally. Second, there is no concept of universal common ancestry in Augustine—species are distinct from each other at their very inception, whether they began to exist fully developed or in seminal reasons. Third, the natural transformation of species, as well as the generation of new natures via accidental changes, is impossible according to the Doctor of Grace.[28] Therefore, in our opinion, Au-

---

28. That Augustine does not allow for the creation of new natures by accidental change, or natural secondary causes, is clear from his treatment of the topic in *The City of God*, III,25 and *On the Trinity*, III,8,15. Since it is not our goal here to fully present Augustine's teachings, we will refrain from providing extensive quotations from his works. A couple of quotations from his *The Literal Meaning of Genesis* explain how he understands the phrase "according to their kinds" from the account of creation: (1) "This then is the significance of *according to their kinds*, where we are to understand both the efficacious force in the seed and the likeness of succeeding generations to

gustine does not support theistic evolution, but excludes it by providing an alternative interpretation of creation. Consequently, the support given to the Augustinian interpretation by theistic evolutionists undermines their own position.

### c. Can Chance Take Part in Creation?

Darwin's theory consists of two factors: chance (random genetic mutations) and necessity (natural selection). The second factor—necessity—is not very controversial because we see necessity everywhere in nature in the form of the laws of nature. The more controversial component of Darwinian theory is the postulate that random biological variations provide sufficient biological novelty to propel natural selection. Moreover, all supposed evolutionary mechanisms currently presented in biology adopt chance as the main driving force. Materialists have been using this fact to counter any idea of design, creation, or purpose in nature. Many Christian evolutionists, on the other hand, have come to the conclusion that the crucial (and for some, the only) problem with evolution is its accidental character. To resolve this problem, theistic evolutionists try to accommodate chance and incorporate it into Christian teaching on creation. As a solution, they present Aquinas's teaching on divine providence.

According to Aquinas, divine providence extends to all beings and all events. It has two aspects: general and particular. This means that both the overall order of the universe and each single—even the least significant—event are known, planned, and supervised by God. But Aquinas recognizes truly random events as well. The problem, therefore, is how to harmonize the randomness of some events with unfailing divine providence. The Angelic Doctor answers that God can bring about His intended effects by using either planned and guided events or truly

---

their predecessors, because none of them was created just to exist once and for all by itself" (*De Genesi ad Litteram*, Book III, no. 19); (2) "We should take [the phrase] *according to their kinds* as in fact meaning according to their species, to distinguish from everything else things that derive their likeness to each other from one original seed" (Book III, no. 20). Both quotations are from *The Literal Meaning of Genesis*, 228. Both fragments demonstrate that Augustine sees the origin of species as the separate creation of the progenitors of each kind, and then their propagation according to the defined limits of their kinds. In the same book he gives a number of examples regarding how to understand words used in Genesis such as cattle, beasts, and reptiles—all of his examples refer to natural species. Hence, he believed that biblical kinds are like natural species, and that they were created as distinct right from the beginning.

chance events.[29] Christian evolutionists adopt the same solution. They say that even though the core of the evolutionary mechanism is random, it does not escape divine government. The interplay of chance and necessity brings about all the biodiversity that is planned and intended by God.

The theistic evolutionists' solution sounds neat and coherent. Perhaps for this reason it has gained overwhelming popularity among Christian scholars.[30] However, there are a few major problems with it.

The first problem comes from Aquinas's explicit teachings on the origin of species. Responding to Avicenna, who believed that species might have been formed by secondary causes, Aquinas says:

> This cannot stand . . . because, according to this opinion, the universality of things would not proceed from the intention of the first agent, but from the concurrence of many active causes; and such an effect we can describe only as being produced by chance. Therefore, the perfection of the universe, which consists of the diversity of things, would thus be a thing of chance, which is impossible (*S.Th.* I,47,1,c).

Aquinas also teaches that accidental differences between individuals cannot produce new species:

> Those things whose distinction from one another is derived from their forms [and this includes different natural species— M.Ch.] are not distinct by chance, although this is perhaps the case with things whose distinction stems from matter. Now, the distinction of species is derived from the form, and the distinction of singulars of the same species is from matter. Therefore, the distinction of things in terms of species cannot be the result of chance; but perhaps the distinction of certain individuals can be the result of chance (*ScG* II,39,3).

We see that for Aquinas there is a difference between generation of an individual belonging to a particular species and the origin of the species itself. Though the former may have some accidental component (such as random genetic mutation), the latter excludes chance. Species were created according to clear ideas in the divine mind; therefore, no chance can take part in the emergence of species. In another place,

---

29. *S.Th.* I,22,4, ad1.

30. This solution was adopted even by the International Theological Commission in *Communion and Stewardship*; see no. 69. For more references on this topic, see my book *Aquinas and Evolution*, specifically chapter 2, which provides more evidence from Aquinas against "God creating through chance."

Aquinas rejects the general evolutionary idea whereby random events play a role in the formation of the universe: "That God acts for an end can also be evident from the fact that the universe is not the result of chance, but is ordered to a good" (*ScG* II,23,6).

When resorting to the idea of chance events being employed by God in creating species, theistic evolutionists confuse creation with generation. Aquinas's teaching about chance events belongs to the order of divine providence. Within this order there is also natural generation, such as a cat begetting a cat and a horse begetting a horse. Accidents (such as accidental mutations) that happen in generation do not escape divine providence, rather they serve God to accomplish his ends. But the origin of species Aquinas attributes to the order of creation. For this reason, Aquinas's teaching on divine providence does not help to save theistic evolution. So far, theistic evolutionists have failed to explain how the origin of species is not the result of chance despite being the product of blind Darwinian evolution.

### d. Theistic Evolution and Historical Realism

From what we said above (III,3,a), it follows that theistic evolution effectively eliminates supernatural formation of the universe. The works of distinction and adornment are replaced by cosmic and biological evolution. From the perspective of theistic evolutionists, the events that Aquinas (along with the entire Christian tradition) ascribes to the works of the six days fall into the seventh day—the day of God's rest and providence. Theistic evolutionists confuse two questions—the question of origins, which has its answer in the biblical history of creation, and the question of providence, which happens now, without transcending the laws of nature (with the exception of miracles).

One of the reasons why theistic evolutionists get mired down in this confusion is that they see nature as a self-assembling entity that can produce ever-growing complexity. They believe that the work of formation has never finished and that the universe is like one gigantic evolutionary process that randomly throws out staggering effects, such as completely new forms of life. And this vision is contrary to our experience, to scientific evidence, to the thought of Thomas Aquinas, and to Christian tradition. Everything tells us that complexity in biology increased only up to a certain point, and after that it has been only diminishing.[31] We do not observe the emergence of new natural species any longer. Instead, nature abounds in extinctions and convergence of races.

Although these facts are not compatible with theistic evolution, they perfectly match the original Christian understanding of creation. Something must account for the initial multimillion-year diversification in biology. According to the doctrine of creation, this is the divine work of formation (*opus formationis*). Considering that nature cannot produce functional complexity by itself, it makes sense that complexity grew only as long as God operated supernaturally in the natural order. After this action ceased (with the creation of man), the natural order started losing all form and structure. Yet, the process of disintegration is slow enough to make salvific events possible in the context of the natural order. In fact, we see that God calibrated the history of salvation in such a way as to match the disorganization of the natural order. If salvation history were situated earlier in natural history, the universe would not have been prepared to provide all the natural resources needed for man. If it were situated later, many natural resources would have been already lost, and many helpful species would have undergone extinction. The work of formation prepared the universe for the history of salvation.

We see that theistic evolution harmonizes evolution with Christianity—but at a high cost, namely, adopting a false vision of nature and compromising the authentic teaching on creation. The deeper problem is that theistic evolution imposes an abstract and a-historical model of the universe that replaces the actual timeline of the creative and salvific events. We see this in the writings of those Thomists who adopt theistic evolution; their work is almost entirely focused on models of divine causality, even though for Aquinas himself the origin of the universe is a matter of

---

31. Surely, there are many examples of the emergence of new biological species (speciation) quoted by Darwin and contemporary evolutionists. But these are not examples countering our thesis, for at least three reasons. First, as Michael Behe has shown, speciation happens primarily owing to the degradation of genomes (breaking genes, inactivating them) and even if it happens owing to mutations, such as duplication, it does not increase the amount of biological information, but rather repeats and swaps what already exists. In any case, these changes do not produce new functional proteins, new organs, or new biological systems. Second, the production of new (biological) species (such as cichlid fish in Lake Victoria) locks populations in their specific environmental niches. This type of adaptation diminishes genetic variety rather than creates it. Third, these examples remain within the level of genera; they do not produce new body plans or new families, and therefore they remain within the range of microevolutionary change, which is not the point of controversy.

history that we know from the Bible.[32] Theistic evolution describes the universe from just one perspective—the current moment of time. The abstract models of divine causality proposed by Thomistic evolutionists replace the actual historical events recounted in the history of creation. This approach resembles the idea of an eternal universe that operates in essentially one evolutionary mode without beginning or end—the idea characteristic of pagan philosophies and mythologies. And this approach greatly differs from the biblical historical realism that speaks about unique events that from time to time greatly change the course of history, a history that is linear—has a first beginning and an ultimate end. This latter approach is characteristic of the Judeo-Christian tradition.

### e. Theistic Evolution and Christianity—A Summary

We said above that there are two Christian interpretative traditions of Genesis. Now, if we consider what is common to both traditions, we can define the essence of the Christian understanding of origins and juxtapose it with the understanding of theistic evolution.

There are three substantial differences between the Christian position and that of theistic evolution. The first concerns the type of causation that brought species into existence. Both Christian traditions support direct supernatural causation, whereas theistic evolution advocates natural secondary causation. The second difference concerns the completion of creation. Both Christian traditions teach that creation was permanently completed with the creation of man, whereas theistic evolution holds that evolution continues indefinitely into the future. Since evolution is a uniform natural process that can constantly generate all kinds of novelties, it follows that the formation of the universe (or the work of creation) has never been completed. At the very least, theistic evolution cannot explain what the Bible means when it says that God rested from His work. It cannot account for the clear distinction in Genesis between the six days and the seventh day. The third difference concerns the origin of species. According to both Christian traditions, species were distinct from their very inception, whereas according to theistic evolution, species share common ancestry, and thus were not initially distinct. Table 3 summarizes these differences.

---

32. Austriaco et al., *Thomistic Evolution*, 65–74.

|  | The origin of species according to **Christianity** | The origin of species according to **theistic evolution** |
|---|---|---|
| Causation | Direct divine | Secondary natural |
| Duration | Finished with the creation of man | Continues as long as evolution operates in nature |
| Origin | Distinct from the beginning | Universal common ancestry |

Table 3. A comparison of Christianity and theistic evolution.

## 4. Christianity and Intelligent Design

### a. The Design Inference and the Design Argument

Having presented the theological challenges to theistic evolution, we move onto a different plane of discussion. Our goal now is to scrutinize the scientific theory of intelligent design from the Christian philosophical perspective.

There is a long tradition in Christianity, confirmed independently by the Bible (Wis 13:1–10, Rom 1:20), that the existence of God can be recognized through the study of nature. The First Vatican Council (1869–1870) made this a solemn Catholic teaching when it proclaimed "that God, the source and end of all things, can be known with certainty from the consideration of created things, by the natural power of human reason."[33] Throughout history, different thinkers, among them pagan philosophers such as Plato and Aristotle, came to this knowledge of God through their own rational investigations. In the thirteenth century, Thomas Aquinas developed the "five ways" that demonstrate how God can be known through the study of nature.

Let's focus for a moment on the fifth way, which will take us to the topic of Christianity and intelligent design. The fifth way begins with the perception of purposeful actions among things that do not possess intelligence. Thomas observes that many things in nature act like an arrow shot toward a target. Since an arrow has no knowledge of where to go, yet reaches its target, there must be someone who gives it direction. Similarly, in nature, inanimate bodies have no knowledge of how to act, yet

---

33. The First Vatican Council, *Dogmatic Constitution on the Catholic Faith*, chapter 2, "On Revelation."

very often they act as if they knew how to accomplish a higher task that serves other creatures or nature as a whole. Aquinas concludes that there must be some intelligent being by whom all natural things are directed to their ends, and this being we call God (*S.Th*.I,2,c). This, along with all similar ways of arguing for God's existence, can be collectively called the *design argument*. The general logical structure of the design argument can be represented by the following syllogism:

1. Everything that is designed must have a designer.
2. The universe is designed; therefore,
3. The designer of the universe exists.

According to William Dembski, the theory of intelligent design does not employ the classical design argument. Instead, it refers to a different type of reasoning that Dembski calls *design inference*.[34] The general logical structure of design inference is represented by a different syllogism:

1. Everything that has certain characteristics indicating design has been actually designed.
2. Some elements in the natural world bear these characteristics; therefore,
3. Some elements of the universe are designed.

We see that the design inference does not take us to God's existence. It stops at the conclusion that at least some natural structures or events are attributable to intelligent causation rather than chance or necessity. The relation between Christian tradition and intelligent design depends greatly on the relation between these two types of reasoning, with their slightly different outcomes.

Both arguments begin with the observation of nature. However, the design argument perceives nature in an abstract way and derives very general conclusions about the totality of things. This approach is typical of philosophical reasoning. And, indeed, only philosophical reasoning can take us from the universe to the very existence of God. Science cannot do this, because science remains entirely within the natural order. The design inference is an argument dealing with particular structures or events observed in nature. It does not enter the level of higher abstraction. It analyzes the particulars in order to draw conclusions about

---

34. Dembski, *The Design Revolution*, 77.

whether the structure or the event is designed or not. This meets the criteria for the scientific type of reasoning. We can say, therefore, that the design inference relates to the design argument in the same way as science relates to philosophy.

Since science remains within the natural order, it is unable to discover God in accordance with its own method. It is clear, therefore, that the First Vatican Council's teaching on the ability of natural reason to discover God cannot be applied to a strictly scientific method. A typical philosophical reasoning, in contrast to the scientific method, takes the human mind to the discovery of the so-called preambles of faith (*praeambula fidei*). These are truths about God and his actions that can be known through natural reason. These include God's existence, his basic attributes (such as omnipotence and omniscience), the dependence of the universe on God, and other such basic truths about God. Still, the preambles of faith need to be supplemented by supernatural revelation, because human reason cannot deduce everything about God based on observing nature. In fact, the most important truths about God (that God is the Trinity, that he was incarnated in Jesus Christ, that he is the Savior, etc.) greatly exceed the abilities of human reason to discover, and therefore need to be revealed in a supernatural way.

The conclusion in favor of intelligent design achieved in science relates to the preambles of faith in the same way as the preambles of faith relate to the supernatural truths of faith. We can see intelligent design as a preamble to the preambles of faith. Hence, there is no contradiction or competition between the design inference and the design argument. Since they prove different things, and employ different methods within different domains of knowledge, we can safely conclude that the design inference can function beside and regardless of the design argument, and vice versa. From a broader perspective, we can also say that intelligent design relates to natural theology in the same way as natural theology relates to divine theology. And since philosophy is a handmaiden of theology (according to the classic adage: *philosophia ancilla theologiae est*), we can also conclude that philosophy finds its handmaiden in the scientific concept of intelligent design. These relations are summarized in Table 4.

| LEVELS OF HUMAN KNOWLEDGE | PARTICULAR DISCIPLINES | PARTICULAR CONCEPTS | ORDER OF DEPENDENCE |
|---|---|---|---|
| THEOLOGY | De Deo Uno et Trino (Treatise on God) | God's existence as stated in the Bible | Regina scientiarum |
| PHILOSOPHY | Natural theology (physico-theology) | design argument | Ancilla theologiae |
| SCIENCE | Theory of intelligent design | design inference | Serva ancillae |

Table 4. The interdependence between levels of knowledge, disciplines of knowledge within the levels of knowledge, and particular concepts within the disciplines.

### b. Intelligent Design and the "God of the Gaps"

In Chapter II (specifically II,5,g–h), we responded to the common charge that intelligent design is not science. We argued that intelligent design can be, indeed should be, included in science, as it neither employs theology/philosophy nor violates the principle of methodological naturalism. At the very least, intelligent design is not less scientific than Darwinism. Here we will respond to a theological objection against intelligent design, i.e., the charge that it is a "god of the gaps" argument.

"God of the gaps" is a popular phrase that refers to a supposedly improper causal explanation of natural phenomena. The god of the gaps argument takes place when someone, after not finding a natural cause for an event, concludes that this event must be caused by a supernatural (and therefore undetectable via the scientific method) power of some kind. The example of primitive people is often presented to illustrate god of the gaps reasoning. For them, many natural phenomena—such as thunder, eclipses, tides, earthquakes, volcanic eruptions—did not have natural explanations. For this reason, they ascribed these events to supernatural causes, such as gods who were angry, or hungry, or anything of this kind.

As a result, nature was generally considered a mysterious entity in which gods, spirits, or demons dwelled and acted according to their needs or whims. It was not even quite clear where nature ended and the supernatural order began. God of the gaps reasoning ultimately led to spiritualization of nature and confusion between natural and supernatural.

Critics of ID think that the same problem applies to intelligent design. They believe that referring to intelligent causality in science is the same as filling the gaps in human knowledge with God or some unknown supernatural power. In addition, some critics of ID are concerned about the possibility that as scientific knowledge expands and fills in the gaps, God will be displaced and we will end up with a worldview in which God has no place at all. To avoid this kind of "delusion," we should accept only natural explanations for natural phenomena. Philosophy and theology on their part should develop ideas about how God works in the universe through secondary causes. Critics think that this is the way to maintain the proper separation of science and theology.

Yet, there is much confusion in rendering intelligent design as a god of the gaps argument. First, we need to notice that primitive people simply did not look for other-than-religious explanations. They did not introduce a god or spirits into gaps in their natural knowledge, because for them explaining the universe was outside the scope of natural knowledge. The religious interpretations of natural phenomena were satisfactory for them. The desire to find the physical causes of physical phenomena was born in ancient Greece and grew in the Christian era. Over the centuries of scientific progress, many natural phenomena did gain natural explanations. But does this mean that God was removed from our knowledge, or perhaps that now we better understand the role of God in the universe? Thanks to science, theology can see more clearly what should be attributed to a supernatural cause and what should not, and thus it gains a better understanding of divine causality. But this is not tantamount to saying that God was removed from our worldview. Today, we know, for example, that God created a uniform universe in which all ordinary operations of nature can be explained by natural laws. Yet, this reasoning does not apply to the origins of the universe. The current operation of the natural order is attributable to natural causation, but this does not mean that the origins of the universe have the natural explanation. We see, therefore, that the progress of civilization (culture, science, theology) does not necessarily remove God from causal explanations but rather helps us better understand which phenomena are properly attributed to either natural or supernatural causes. And this is different from the idea of one-directional progress consisting of removing God from causal explanations.

The second problem involves the principle of methodological naturalism. If we want to avoid the god of the gaps in the way the critics of ID present it, we should give up any supernatural explanations by default, even before we conduct any inquiry. Surely, then, we would avoid the god of the gaps argument. But would it warrant true explanations? It seems much more reasonable to say that science (as well as philosophy and theology) should be guided by the principle of seeking the *best* rather than only the *best natural* explanations. And this principle remains valid even if in science a natural explanation happens to be better. It does not follow that scientific inquiry should violate the principle of methodological naturalism. It just means that science needs to be able to see its own limits and admit that some events cannot be explained by its method. Science cannot be satisfied with a natural explanation that does not explain anything. A good example is a miracle. In this case, scientists should be able to suspend the scientific method and acknowledge that there is something beyond the competence of science.

Third, as we explained above (see II,5,h), in our opinion intelligent design does not violate the principle of methodological naturalism. Hence, ID does not introduce a god or any supernaturality into science. Intelligent design only supplements the array of causes recognized in science. So far only two have been widely recognized—necessity and chance. Intelligent design says that a third type should be added, that is, intelligence. This is analogous to a less controversial claim regarding the existence of three factors in the formation and operation of the natural universe. Until recently, only two of them were universally accepted, namely, matter and energy. Today, hardly anyone denies that there is also a third element—information. The controversy is not so much about whether information exists in the universe or not, but where it comes from.

In fact, many scientific disciplines depend on our ability to infer design. For instance, in archeology it is necessary to decide whether some small rocks are tools intentionally made—stone axes, arrowheads, knives—or merely the result of erosion. In forensic science, a detective needs to decide whether the event in question was random (as with an accidental death) or intentionally planned (as with a murder). In computer science, engineers may need to establish whether a string of digits is random or a code, a program. When a patent office suspects intellectual theft, again, it is necessary to establish whether the similarity between the designs is accidental or intentional. In all of these cases we deal with science, which implies that inferring design in science is not only possible but even common. Since the detection of design works in many branches of science, there is no reason to bar it from biology.

Fourth, the god of the gaps argument does not apply to ID because ID is based on knowledge rather than ignorance. The argument says that we insert a god into the gaps in our knowledge about nature whenever we do not find a natural explanation. But ID is based on thorough study of natural structures and events that leads to a conclusion that we need an intelligent cause. It is not supplementing ignorance; it is positive knowledge supported by evidence. Indeed, the "explanatory filter" (see II,5,b) first looks for explanations from chance and natural laws, and only after finding that neither can account for a given structure or event, it resorts to intelligent causation. Similarly, in order to establish irreducible complexity, one needs to know all the parts of a system and what role they play. Only after we know this, may the conclusion that all of them are necessary to perform the basic function of the system be permitted. The logic of the ID argument is exactly opposite to the logic of the god of the gaps. The latter assumes supernatural where it does not find natural, whereas the former affirms intelligence where it excludes chance and necessity while discovering purposeful design.

### c. Intelligent Design and Philosophy of Nature

It seems that no modern idea has challenged scientific materialism as effectively as intelligent design has. One may wonder, then, how it is possible that many traditionally minded believers, people who generally have good intentions, understand classical philosophy, and want to oppose atheism and materialism, would ally with the atheists and materialists in opposing intelligent design? Why does this theory encounter resistance from, for instance, Catholic philosophers of nature?

To answer this question, we need to find the root causes, and to do this we need to dig deep. It seems that one of the causes is the unstable status of so-called "philosophy of nature" in our times. Philosophy of nature was the way of knowing the natural world in antiquity and the Middle Ages. It was a legitimate and even influential (if not predominant) part of philosophy before the scientific revolution. Its goal was to explain nature in philosophical terms, i.e., to draw general conclusions about physical reality through abstract and deductive reasoning, observations of uniform movements, and so on. The best example of this approach is Aristotle's *Physics*, which laid the foundations for this entire branch of philosophy for many centuries.

However, when in modernity many exact sciences branched out from philosophy of nature, tension appeared. Philosophers wanted to maintain their privileged position as those "who know the truth" and govern other disciplines. At the same time, modern scientists, by employing the experimental method and mathematical description (while rejecting metaphysical, abstract reasoning), proved the philosophers wrong regarding many natural things. The controversy over Galileo's scientific proposals exemplifies this problem quite well. Moreover, scientists became extremely successful not just in explaining physical phenomena but also in making people's lives easier by means of technology and innovation. Science had an overwhelming practical aspect, which was almost completely missing in philosophy of nature. Consequently, science attracted more attention and eventually gained higher priority in the popular reception of these two domains in culture. A conflict arose between philosophers of nature and scientists. In reaction to the predominance of the nominalistic and pragmatic approach among scientists, some philosophers downplayed the role and "veracity" of modern science as such. Some even claimed that science is just a collection of ever-changing, superficial postulates that never provide any true knowledge. In contrast, they claimed, philosophy is the discipline that gives an *understanding* of nature and yields permanent and true insight into natural phenomena. Scientists, for their part (if they were at all bothered by any of this criticism), usually equated philosophy with "scholasticism" and presented it as "intellectual fiction," an outdated way of thinking that is even detrimental to culture because it slows down human progress (see our account of Comte's historiosophy in I,3,a). Positivistic scientism deemed philosophy a fairy tale and deprived it of the status of knowledge.

This old and ongoing conflict between philosophers and scientists partly explains the contemporary attitude of doubt and reluctance exhibited by some philosophers with regard to intelligent design. However, there are two other main factors at work here.

The first is the fact that intelligent design was born among scientists and not philosophers. Classically minded philosophers, who have been licking their wounds after two centuries of serious defeats from secular culture (and especially the scientism of that culture), would never expect any good to come directly from science. So, their first reaction is surprise and incredulity with respect to the abilities of contemporary science. They seem to react like those asking in Jesus's times: "Can anything good come from Nazareth?" (Jn 1:46).

This, however, is followed by the second factor, which is an attitude of suspicion and fear that science would appropriate the competencies of philosophy. As long as naturalistically and nominalistically minded scientists say that there is no philosophy whatsoever, their claims are extreme enough to be not seriously considered by a broader audience. Indeed, presenting any idea, even one as popular as scientism or materialism, in an extreme way does not do good for the idea itself. Philosophers and theologians do not deem dangerous such extreme opinions as "philosophy is like mythology" or "God is dead." This may also explain why philosophers find it easy, or maybe even entertaining, to debate extremists like Richard Dawkins or Daniel Dennett. But the idea that science can detect design seems to philosophers both bold and dangerous. They see ID as competition—a tool for finally defeating philosophy. What scientism has not achieved by the total rejection of philosophy it may achieve by incorporating a little bit of philosophy into science. This "little bit" is exactly what justifies the very existence of philosophy of nature in our times. Hence, if this were removed from philosophy and placed in the realm of science, there would be no reason to pursue philosophy of nature. Philosophers of nature are tempted to think along the following (or similar) lines: "As long as science can provide only a purely materialistic worldview, philosophy plays a crucial role in culture, because it is the only alternative to naturalism. Philosophers are the only ones who can save the Christian worldview. However, if science is capable of discovering intelligent design, then it can also speak about teleology, purpose, etc. All of this belongs to the philosophical domain. This is entering our field and taking our jobs!"

Philosophers of nature know that they don't have much to offer to today's science in terms of explaining natural phenomena. Instead, they see their role as mentors who encourage the less pragmatic part of society to think in more abstract terms, to recognize things like the sensory and vegetative souls in animals and plants. According to them, finding design in nature also belongs to that "deeper" reflection, that is, to philosophy. For this reason, in extreme cases philosophers of nature see ID as a cunning tactic used by the forces of scientism to destroy their domain. In less extreme (and more common) attitude, they see it as materialistic reductionism that looks for design without any reference to metaphysical notions and principles. At the very least, they consider ID a poor philosophy that must be corrected by "true" philosophy, that is, philosophy of nature.

Based on what we said above about the relations between science and ID (II,5,g–h) and Christianity and ID (III,4,a), we can conclude that philosophers do not have a valid reason to be suspicious or resentful of intelligent design. Surely, there will always be scientists who would like to appropriate some of the philosophical domain, but there are also philosophers who tend to replace natural science with philosophy. Misunderstanding or even disrespect may arise on both sides, but this is not how science or philosophy works in the first place. Intelligent design is an empirical concept and, by itself, does not reach for any stronger philosophical claims. Hence, if philosophers of nature recognized the truly scientific status of ID, they would not be afraid of competition between ID and philosophy of nature.

Moreover, ID in science does not endanger the autonomy of philosophy. Even if science may recognize that some structures or events should be attributed to intelligent rather than random or law-like causation, philosophy of nature remains a fully autonomous discipline concerned with its own issues. The entire Aristotelian theory of four causes, as well as any "deeper," more abstract, or more insightful approach to nature, remains valid on its own terms. Intelligent design is a link that apparently is missing in the contemporary debates between "scientific atheists" (such as Dawkins) and Catholic philosophers of nature.

Additionally, ID helps to reconnect philosophy with science. Before the theory of ID was created, atheists had an easy task: they simply advocated the existence of two domains. One, they said, was the domain of natural sciences, and this was empirical, certain, and objective; the other was the domain of philosophy, theology, mythology, and other "fictions," which had some historical and cultural merit, but did not provide any useful knowledge and should not in any way shape our worldview. Intelligent design breaks this logic at the very core by showing that science itself is open to, indeed encourages, the asking of deeper metaphysical questions.

The atheists' approach is based on the sharp division between the two domains—the natural and the supernatural—and on the disdain often shown for the latter. Surprisingly enough, some classically minded philosophers adopt a similar attitude. The only difference is that for them philosophy is what reveals the important message about the universe and science is what oscillates between phenomenology and mythology. The antisynthetic approach of these philosophers separates them from the true Christian tradition, even if they claim to be its custodians. Christian tradition has no problems with truly scientific progress and even with accommodating and adopting whatever good can be found in science and culture.

Moreover, it should not be surprising for any Christian that God uses even such insignificant domains as natural science to make people think about him. After all, as Christianity sees it, all being, truth, and good come from one source, who is God the Creator. Intelligent design is a scientific theory that enables Christians to rebuild the science-faith synthesis that has been ruined in modernity by unwilled but difficult-to-avoid conflicts between science and religion. We can say that intelligent design is the missing link in the current debate between Christianity and the scientism and technocratic reductionism of our times; it remains entirely within the domain of science, yet it enables a smooth transition to philosophy and theology.

### d. What's the Problem with Philosophy of Nature?

We said that philosophy of nature (or natural philosophy) entertains just a modicum of respect in modern academia. One reason for this downfall is that philosophers do not clearly distinguish or even allow for a distinction between natural science and philosophy. In earlier parts of the book (I,1 and 3), we tried to thoroughly define these differences according to the object (subject matter), the method, and the goal of each discipline. Now we want to seek for a deeper problem that resides at the root of the modern conflict between classically minded philosophers and scientists. The problem pertains to the status and the character of philosophy of nature.

Ancient philosophers wanted to explain all of reality, including both the visible and invisible realms. Their approach to the universe was prereflective, i.e., it followed the natural attitude of the human to external reality. This kind of approach in philosophy is called moderate realism, the belief that there is an external world outside our mind that we, thinking subjects, can apprehend by our senses.

But ancient philosophers were not simply satisfied with perceiving nature. They wanted to explain the ultimate causes of the universe. In order to do it, they had to transcend the material perceptions in their thinking. Since they lacked supernatural revelation, the only way for them to do so was by producing more and more abstract notions that originally stemmed from that natural experience of the surrounding material universe. This highly abstract system of thought, derived from material reality, was the greatest achievement of ancient philosophy because it elevated the human mind to the contemplation of the first being, the Absolute. The way to come to these highest conclusions runs through a philosophical discipline called metaphysics.

The name "metaphysics" means "after physics" and derives from the first complete edition of Aristotle's writings, in which the work *Metaphysics* was placed after another work, *Physics*. The order of publication coincides with the content. Whereas in *Physics* Aristotle wanted to explain the natural world (the Greek word *phusikos* means natural), in *Metaphysics* his goal was to explain the invisible things and specifically the first being, whom we call God. We can see therefore a directed movement in Aristotle's works: from the natural experience of the universe to the search for the general principles and explanations of nature. But explaining nature is not the last step; there is a further journey that takes the mind by means of abstraction toward the most general and most fundamental principles of all reality which bring us to contemplation of being *per se*. How does it connect with the problem of philosophy of nature?

First of all, we should observe that Aristotle pondered upon nature not to create a "philosophy of nature" but to explain operations of nature. His goal was as pragmatic as that of modern science—to explain why material things act the way they act. But his knowledge of nature was phenomenal, because it was based on simple, everyday perceptions. Aristotle was aware that this kind of "natural phenomenological knowledge" was not sufficient to explain the universe, so he started observing nature in a more systematic way.[35] He wanted to gain a better, more exact, understanding of what actually happens in the physical realm. But his experimental knowledge was very limited. He did not know how nature operated; he only knew how it appeared to the senses. One way to supplement this lack of experimental knowledge was to use general abstract principles (that were useful in metaphysics) applied back to nature. This

---

35. Aristotle did not perform experiments in the modern sense; however, he did some more insightful and rigorous observations (Greek *pepeiramenoi*). One example is finding a fertilized hen's egg of a suitable stage and opening it so as to be able to see the embryo's heart inside.

led to conclusions about nature, such as that the planets should follow the path of perfect circles because it is the best trajectory of motion, or that everything tends toward its natural place (for example, fire tends to move upwards and earth downwards). Aristotle supplemented his phenomenal knowledge of nature with a deductive mode of explanation.

This kind of "physics," namely phenomenal and deductive, was coherent and successful at explaining the universe until novel experiments beginning with early modernity brought forth new knowledge. All of a sudden it turned out that phenomenal knowledge was not always "true knowledge." New observations and experiments showed that nature does not work the way Aristotle imagined, and thus that his explanations were not always accurate. Over time the discrepancy between the "scientific" and "philosophical" explanations of nature increased, creating tension and conflict between modern science and natural philosophy. The Galileo affair is an example *par excellence* of this modern problem. It seems that within a few centuries modern science won the conflict and superseded Aristotelian physics in explaining the universe. Does this mean that philosophy of nature has been nullified by modern science?

In order to answer this question, we need to acknowledge that the ancients did not have a good grasp of natural phenomena and that for this reason their explanations of nature were often incorrect. We can say that they were "phenomenologists of nature," in the sense that they did not conceive the material things as they *are*, but only as they *appear*. By the same token, modern science is "realistic" in the sense that it takes us to the true causes of natural phenomena.

This, however, does not mean that classical philosophy should be abandoned. Abstract notions employed in metaphysics, such as substance, form, matter, act, potency, etc., do not change over time, because they describe reality on the highest level of abstraction. At this level, reality remains unchangeable throughout all times and metaphysics remains as valid today as it was two millennia ago. However, philosophy of nature operates on the lower levels of abstraction, which makes it dependent on new discoveries and necessitates its actualization according to the new knowledge about the universe. It seems, therefore, that today philosophy of nature is either reducible to science or serves merely as a somewhat more abstract speculation based on scientific theories that are at the edge of science, such as those in cosmology or quantum physics.

There is a little bit of irony in the fact that the ancients, who did not have a realistic view of nature, created the realistic philosophy, whereas we, who have developed a more realistic knowledge of nature, tend to embrace idealistic philosophies. We can even say that the ancients were phenomenologists and idealists regarding physics, but realists in metaphysics, whereas modern people are realists in physics, but as if at the cost of abandoning realism in metaphysics.

The bottom line is that sound philosophy (*sana philosophia*) consists primarily of metaphysics, which should be defended against all variants of reductionism, and propagated as the philosophy whose explanatory power has not diminished since antiquity. Indeed, no other discipline can explain reality, whether visible or invisible, as insightfully as metaphysics. This is also why it is called *philosophia perennis* (eternal philosophy): it does not change with the progress of science. Philosophy of nature, on the other hand, is sensitive to the progress of natural science, and must be constantly updated according to the growing body of knowledge about nature.

The way to overcome the long-lasting conflict between philosophy of nature and modern science is to make two clear distinctions. The first concerns the limits of natural science and philosophy. As long as their proper domains are not recognized and science is called philosophy, or philosophy is reduced to science, the conflict will last. The second distinction must be recognized within philosophy itself, that is, the distinction between physics (in the Aristotelian sense) and metaphysics.

In philosophy, the truths and principles that are permanent and unchangeable (owing to their abstract character) must be distinguished from those that are less abstract, more connected to nature, and changeable according to new discoveries in science. Only the permanent elements (metaphysics) deserve philosophical defense against new scientific theories (if those theories challenge metaphysics). Philosophy of nature relying on inadequate perception of nature may be wrong in its explanation of natural phenomena. But metaphysics relies on permanent notions and principles that touch the very essence of reality, so it cannot be overturned by any particular theory proposed in natural science.

If a scientific theory challenges classical metaphysics (as in the case of biological macroevolution), then we have good reason to suspect that the theory is false. We should disprove it by employing higher, metaphysical principles. But if there is a new discovery in science, such as intelligent design, that is not in conflict with metaphysics, it should be acknowledged by philosophy of nature so that philosophy of nature does not conflict with modern science.

### e. Philosophy of Nature and Biological Macroevolution

Many contemporary philosophers of nature believe that, while biological macroevolution (or neo-Darwinism) can be reconciled with traditional natural philosophy, the theory of intelligent design cannot.[36] Some of the reasons why this paradoxical situation predominates in Catholic academia were already explained (in III,4,c). Here we would like to highlight one particular, but very telling, aspect of this inconsistency in the Christian philosopher's worldview.

Traditionally minded philosophers often lament the reductionism and materialism of natural sciences. Their strategy is to fight reductionism by popularizing Aristotle's concepts pertaining to nature and encouraging scientists to open their minds to the philosophical interpretation of physical phenomena. One significant element of such a "non-reductionist biology" is the appeal to the immaterial principle or the "form" of a living being. According to Aristotle, every living being has an immaterial element that organizes the body (the material element) in such a way that the body performs functions otherwise inaccessible to purely physicochemical entities. This is clear when we consider the many examples of generation and regeneration in animal life. No mechanism, whether physical or chemical, can function in the way living beings do. Machines cannot repair themselves, and they cannot reproduce. Of course, machines can produce entities that are similar to each other, but very different from the parental machine, as happens in factories. For example, a fully automated car production line is a machine that makes other machines (cars) without any guidance from an intellect or an immaterial form. However, it is not that a car makes a car, as a dog generates another dog. A factory is a much greater and much more complex machine that is capable of repetitively producing much simpler machines. Machines do not produce things like themselves; neither can they fix themselves, whereas living beings do so. This is one of the proofs that in biology we do not deal with merely biochemical mechanisms. The difference between physical and biological structures is not just quantitative, as if it were a matter of merely a more complex arrangement of particles, but also qualitative. Living beings have another level of organization of matter which is produced by the animal soul, which gives matter the ability

---

36. Currently quite a few scholars, including many Thomists, advocate this position: Edward Feser, Marie George, William Carroll, Francis Beckwith, Michael Tkacz, and Gerard Verschuuren, among others.

to act beyond what is available to matter on its own. A living being is a complete, functioning, and coherent entity that is not reducible to its parts. The rejection of this ontological difference by treating living beings as machines (in an absolute sense) is exactly the type of reductionism that bothers philosophers of nature.

But here is the problem: The same philosophers who combat this kind of reductionism accept the core neo-Darwinian postulate that living beings were produced through natural, biological processes that function like mechanisms. A mechanism can produce only a mechanism (as a car factory produces cars), but never a living being with an immaterial principle—the soul. Philosophers of nature who believe in species being produced by evolution fall into the same type of reductionism they criticize among biologists.

Allowing for the special creation of the first cell (so that the immaterial principle is present in the first living organism) does not really help to overcome this conflict. If diversification into many species happened owing to a biological mechanism it would still imply that new forms (immaterial principles) can be produced by manipulating the material substrate of living beings.[37] The conclusion that matter can be the active principle creating new substantial forms is contrary to the Aristotelian-Thomistic approach.[38] And it does not even matter whether the biological evolutionary process would be guided or random. Theistic evolutionists

37. This is a solution to the conflict between classical philosophy of nature and biological macroevolution proposed by, among others, Charles De Koninck. See his "The Cosmos: The Philosophic Point of View," 256–321, 278–83. Michael Bolin pushes this argument to its limits when he proposes that once matter achieves the proper disposition, God directly induces a new form into the being. Bolin's position should be properly called "materialistic occasionalism." See Bolin, "And Man Became a Living Being."

38. There is a large body of evidence for this claim. For example, Thomas Aquinas speaks about the relationship between matter and form: "It is clear that something is in act according to the form, according to matter is in potency and according to the dispositions of matter is apt to the act" (*Super Sent.* lib.4, d.49, q.3, a.2, co). "Matter is for the sake of the form, and not the form for the matter, and the distinction of things comes from their proper forms" (*S.Th.* I,47,1, co). And in another place: "Forms are not consequent upon the disposition of matter as their first cause; on the contrary, the reason why matters are disposed in such and such ways is that there might be forms of such and such kinds. Now, it is by their forms that things are distinguished into species. Therefore, it is not in the diversity of matter that the first cause of the distinction of things is to be found" (*ScG* II,40,3). Aquinas repeatedly says that new complete natures cannot be produced by a change, let alone a physical process or an accumulation of accidental events. For more evidence see our book *Aquinas and Evolution*, especially chapter 2.

believe it is guided by the divine mind, though they can hardly explain how it happens (i.e., how God can guide an unguided process). But the crux of the problem is not that evolution is unguided, blind or random (as atheists say), but that it is a process, i.e., interaction of physical particles and bodies. No material process, no matter how complex, is capable of producing a non-material principle which would organize matter in such a way as to produce a new species.

We can also approach this issue from a different perspective. Evolution as a biological process acts by affecting parts of living things. We can list thousands of such changes that biological processes produce in microevolution, but they always affect just a given part, or a given structure in a living thing. For example, genetic mutations affect some nucleotides in the DNA, which can lead to modified proteins, or a modified body structure. In any event, we always talk about one or many parts of a living being that are modified by a biological process. But if a process is supposed to produce new living beings by merely modifying parts of old living beings, it means that the new living beings are nothing more than the conglomerates of the parts. And it does not matter how many parts would be physically modified. According to classical philosophy of nature, even if all and every single atom in the body were modified by a process, still no new immaterial form (or a new nature) would be produced.[39] Claiming otherwise boils down to adopting a type of materialistic reductionism in which matter dominates over form and produces forms.

We see that any evolutionary scenario in which a material process (guided or unguided) is supposed to generate species is a type of reductionism that clashes head-on with the principles of the classical philosophy of nature. It is therefore strikingly incoherent to allow natural macroevolution on the one hand and fight biological reductionism on the other. Even so, many contemporary philosophers of nature fall into this contradiction because they do not consistently adhere to the principles of classical philosophy and accept evolution under the pressure of the "scientific community." But scientists who make claims about the origin of species are not, strictly speaking, scientists; rather, they speak as philosophers of nature. Those philosophers of nature who believe in

---

39. This is also why it is impossible to produce life from non-living matter. Theoretically, we could arrange all elements and compounds as they are in a living being, but it would not live, because the immaterial principle of life would be missing. We would obtain only dead matter, meat, flesh, the organic substrate of life, but not a living organism.

biological macroevolution unfortunately fail to separate science produced by scientists from poor philosophy produced by scientists with materialistic agenda.

## 5. Catholicism and Evolution

### a. Has the Church Ever Condemned Darwin?

In contemporary Catholic scholarship, it is a popular trend to defend theistic evolution by referring to the authority of the Church. Theistic evolutionists often say that the works of Charles Darwin have never been condemned by the Church, and therefore evolution is acceptable to a Catholic. Leaving aside the non-sequitur aspect of this argument, we need to ask if the main premise is actually true. Surely, if the teaching office of the Church has never condemned Darwin, belief in evolution would be easier to accept for Catholics. However, Darwin advocated not just the idea of biological macroevolution but also a particular mechanism intended to explain it. Theistic evolutionists sometimes acknowledge the fact that the Darwinian mechanism cannot really account for the origin of species, or that the random nature of this mechanism is irreconcilable with Christianity. To overcome this problem, they distinguish between the "Darwinian mechanism" and "evolution itself" (understood as biological macroevolution), and then maintain that evolution can be true even if the Darwinian mechanism is scientifically or theologically untenable.[40] This puts them in quite a convenient position, because they do not say that biological macroevolution is true, but merely biologically "possible" and theologically "acceptable." However, if anyone challenges either the "possibility" or the "acceptability" of biological macroevolution,

---

40. The first Catholic evolutionists already made the distinction between the "mechanism" of evolution and the "fact" of evolution. St. George Mivart explicitly rejected the mechanism of random variation and natural selection; this, however, did not deter him from accepting fully-fledged theistic evolutionism. Similarly, Rev. Henri Dorlodot and his student, Rev. Ernest Messenger, justified the overall evolutionary paradigm on theological grounds regardless of possible limitations of the evolutionary mechanism as presented in science. See Mivart, *On the Genesis of Species*; Dorlodot, *Darwinism and Catholic Thought*; Messenger, *Evolution and Theology*. No less characteristic is the attitude of John A. Zahm, who, after analyzing all the mechanisms of evolution proposed in his time, finds that none is sufficient to explain all the facts. Nevertheless, Zahm concludes: "Whatever may be said of Lamarckism, Darwinism and other theories of Evolution, the fact of Evolution, as the evidence now stands, is scarcely any longer a matter of controversy" (Zahm, *Evolution and Dogma*, 201).

the same scholars who normally seem somewhat indifferent immediately turn into ardent defenders of the whole evolutionary paradigm. Hence, even though Catholic evolutionists very rarely openly advocate Darwinism, they strongly defend the evolutionary paradigm in theology. This indicates that the theological implications of Darwinian theory are more important in the ecclesiastical debates than Darwinian science.

Since the Church is competent in judging theological matters, it is important to know what the actual stance of the Catholic Church on evolution is. In the book *Catholicism and Evolution*, we presented in greater detail the history of Catholic teaching on evolution, without downplaying the initial rejection of the theory or exaggerating the contemporary implicit acceptance of the theistic form of evolution.

To answer the question about the "condemnation of Darwin," first we need to see that a given concept (let's call it concept A) may be rejected in two ways: either explicitly, by saying "concept A is wrong," or implicitly, by proposing in a positive way another concept that excludes A. For example, when one says, "Peter is not going to the cinema tonight," one explicitly excludes Peter's presence at the cinema tonight. If one says, "Peter is working at home tonight," one excludes Peter's visit to the cinema, though not in a positive or explicit way, but implicitly.

Now, the idea of Darwinian evolution was proposed in the name of science and is commonly recognized as a scientific idea, regardless of how scientific it actually is. The teaching office of the Church exhibits infallible judgment in matters of faith and morality, but not in matters of science. Although it is not impossible, in principle, for the Church to make judgments regarding scientific theories, the Church is reluctant to do so because she wishes to avoid errors and remain focused on her essential mission. Therefore, there is nothing definitive in the fact that the Church has never made any universal declaration condemning Darwin. However, we need to notice that even though there has not been a public condemnation of Darwin's theory, neither has the Church embraced or commended his ideas. Thus, the argument from "non-condemnation" of "Darwin's science" is somewhat weak, because it does not say much about the Church's position regarding the theological implications of Darwinian theory.

Having said that, we need to add that there was actually at least one private condemnation of Darwinism explicitly addressing the Darwinian system and mentioning Darwin by name. It was issued by the Congregation of the Index, and accepted by an independent decision of Pope Leo

XIII. This document, even if hidden in the Vatican Archives, is a valid decision representing the will of one of the institutions representing the teaching office of the Church. The decree was issued in 1878 against a book by an Italian priest, Raffaello Caverni, who attempted to combine Catholicism and Darwinism. The relevant portion of the decree reads:

> [Caverni's book] merits serious and special attention. In it, Darwinism is expounded and partly approved, [stating] that it has many points of contact with religious doctrine, especially with Genesis and other books of the Bible. Until now the Holy See has rendered no decision on the system mentioned. Therefore, if Caverni's work is condemned, as it should be, Darwinism would be indirectly condemned . . . With his system, Darwin destroys the bases of revelation and openly teaches pantheism and abject materialism. Thus, an indirect condemnation of Darwin is not only useful, but even necessary, together with that of Caverni, his defender and propagator among Italian youth.[41]

It is true that this document's low degree of authority and private character do not allow it to be used as a decisive tool in a possible argument against evolution. Nevertheless, this document renders equally unfounded all claims about the supposedly peaceful, neutral, or positive acceptance of Darwinian theory within the first decades after *The Origin of Species*. The Congregation of the Index adopted a coherent and explicit stance against any attempts to incorporate Darwin's theory into Catholicism. In addition to Caverni's, also books by Fathers Dalmase Leroy and John A. Zahm were suspended. None of them advocated atheistic evolution or the evolutionary origin of the human soul. These authors supported theistic evolution and the special creation of the human soul, and even though two of them (Caverni and Leroy) excluded the human body from evolution, their work was still rejected by the Congregation. In light of these facts, there is little doubt that the Church, represented by one of her congregations, initially recognized theistic evolution as a dangerous modification of traditional belief, falling outside of what could be considered permissible in the context of Catholic orthodoxy.

If Darwin's theory is scientific, then, as we said, judging it in a direct way does not fall within the primary mission of the Church. However, the Church may reject Darwin's theory by presenting an alternative doctrine regarding origins. The Church has never been deeply concerned

---

41. Quotation from Artigas et al., *Negotiating Darwin: The Vatican Confronts Evolution*, 47.

with the origin of animals and plants, leaving the question to the judgment of sound philosophy. Yet, when it comes to the origin of man, the Church adhered to an explicit and well-established teaching based on the Bible and sacred tradition which was still maintained for nearly a hundred years after the publication of *The Origin of Species*. We can therefore conclude that the Church rejected the evolutionary origin of man, not directly but implicitly, by putting forward a differing positive doctrine on the special creation of both man's soul and body. Since the human body was created directly by God, it could not have evolved naturally, even if the supposed evolutionary process was guided or inspired by God. Special creation excludes natural evolution.

The direct formation of the first human body from the dust of the earth was considered Catholic teaching from antiquity until Pope Pius XII's encyclical *Humani Generis* (1950). In fact, there is overwhelming evidence from Church Fathers, Doctors of the Church, medieval saints, and the common agreement of theologians testifying to the literal understanding of the creation of man from the dust of the earth.[42] As early as the sixth century, Pope Pelagius I proclaimed in his solemn confession of faith that

> Adam himself and his wife were not born of other parents, but were created: one from the earth and the other from the side of the man.[43]

Shortly after Darwin came up with his theory, the local ecclesiastical synod in Cologne (1860) made the following judgment:

> Our first parents were created immediately by God. Therefore we declare that the opinion of those who do not fear to assert that this human being, man as regards his body, emerged finally from the spontaneous continuous change of imperfect nature to the more perfect, is clearly opposed to Sacred Scripture and to the Faith.[44]

---

42. See the appendix to *Catholicism and Evolution*, 310–335.

43. The Pope first pronounced it in a letter *Humani generis* to king Childebert I (of 3 February 557). The entire formula reads: "I confess . . . that all men from Adam onward who have been born and have died up to the end of the world will then rise again and stand before the judgement-seat of Christ, together with Adam himself and his wife, who were not born of other parents, but were created: one from the earth and the other from the side of the man." See Denzinger–Schönmetzer, *Enchiridion Symbolorum Definitionum et Declarationum*, ed. 34, Herder 1967, 155. DS 443 (228a).

44. Provinciae Coloniensis, *Acta et Decreta*, 30.

The decrees of the synod in Cologne were independently recognized by Pope Pius IX. Twenty years later, Pope Leo XIII confirmed the same teaching on human origins in his encyclical *Arcanum Divinae Sapientiae* (1880). Another few decades later (1909), the Pontifical Biblical Commission established that the special creation of man (*peculiaris creatio hominis*) and the formation of the first woman out of the first man (*formatio primae mulieris ex primo homine*) cannot be doubted as *historical* and *literal* truths contained in Genesis. The same is taught by the first Catechism of the Catholic Church (1566) when it says that God "formed man from the dust of the earth, so created and constituted in body as to be immortal and impassible." The same doctrine was present in all local catechisms (if they addressed the problem of the origin of man), of which the Baltimore Catechism, No. 4, is the most explicit: "On the sixth day God created man and called him Adam . . . God could have made Eve as He made Adam, by forming her body out of the clay of the earth and breathing into it a soul, but He made Eve out of Adam's rib to show that they were to be husband and wife."[45] The same perspective was still present in a 1941 address of Pope Pius XII stating that "only from man could there come another man who would then call him father and ancestor,"[46] which excludes the possibility of man descending from a nonhuman ancestor.

This uninterrupted doctrine was weakened or implicitly abandoned by the same Pius XII, in his 1950 encyclical *Humani Generis*:

> the Church does not forbid that . . . research and discussions . . . take place with regard to the doctrine of evolution, in as far as it inquires into the origin of the human body as coming from preexistent and living matter.[47]

No doubt, different interpretations can be applied to Pius XII's statement. For instance, the pope might have had in mind only that the doctrine of evolution should be studied more thoroughly to expose its errors and make Catholic teaching even more explicit.[48] Nevertheless,

---

45. Kinkead, Baltimore Catechism No. 4, listed in Bibliography.

46. Address of Pope Pius XII to the Pontifical Academy of Sciences, November 30, 1941, 92.

47. *Humani Generis*, no. 36.

48. This interpretation is actually implied by another fragment of the document (quoted already in II,4): "Catholic theologians and philosophers, whose grave duty it is to defend natural and supernatural truth and instill it in the hearts of men, cannot afford to ignore or neglect these more or less erroneous opinions. Rather they must

the fact of the matter is that the encyclical of Pius XII is the first Church document that is not clear about the special creation of the human body.

After *Humani Generis*, theistic evolution kept pervading theology and gradually dominated Catholic education. Catholic apologists, who until 1950 had resisted the implementation of theistic evolution into Catholic scholarship, were deprived of their basic argument—an uninterrupted Church tradition along with the Church teaching office that supported and defended the special creation of the human body. After 1950, we do not find any significant magisterial documents or statements on evolution. A few references made by later popes add hardly anything to the state of the discussion: Paul VI, in one of his speeches, suggested to Catholic scholars that they abandon polygenism.[49] John Paul II said that evolution does not, in principle, contradict belief in creation[50] and that it is something more than a hypothesis.[51] On the other hand, Benedict XVI warned that evolution violates the limits of science and has become a kind of "*philosophia prima*" of our times.[52] However, none of the recent popes clearly defines the term "evolution," let alone makes a doctrinal judgment.

We see that so far *Humani Generis* remains the last significant document on the matter of evolution. It has a doctrinal character, and it explicitly teaches that it is permissible for Catholic scholars to study the evolutionary origin of the human body. Note that the encyclical does not even speak about evolution as such, but only about the origin of man, which is just one issue involved in the broader problem of evolution and creation. Apparently, the Church is either confused on the

---

come to understand these same theories well, both because diseases are not properly treated unless they are rightly diagnosed, and because sometimes even in these false theories a certain amount of truth is contained, and, finally, because these theories provoke more subtle discussion and evaluation of philosophical and theological truths." See *Humani Generis*, no 9.

49. Paul VI said at a symposium devoted to the topic of original sin: "It is evident that you will not consider as reconcilable with the authentic Catholic doctrine those explanations of original sin, given by some modern authors, which start from the presupposition of polygenism, which is not proved" (AAS, 58 (1966), s. 654).

50. "Indeed, the theory of natural evolution, understood in a sense that does not exclude divine causality, is not in principle opposed to the truth about the creation of the visible world, as presented in the Book of Genesis." Catechesis "In Creation God Calls the World into Existence from Nothingness" (29 Jan 1986).

51. Address to the members of the Pontifical Academy of Sciences of 22 October 1996 in relation to the session "The origin and evolution of life," AAS, 89 (1997), 188.

52. See Horn and Wiedenhofer, eds., *Creation and Evolution: A Conference with Pope Benedict XVI*, 18.

matter or waiting for more evidence. Today no party in the Church can claim that their position is that of the Church's magisterium—neither theistic evolutionists, nor progressive creationists, nor young-earth creationists. It is possible, however, that in the future the Church will make a definitive judgment.

## b. Faith and Science in the Current Ecclesiastical Debate

Based on what we have presented thus far, we can now sketch a map of ideas currently present in Catholicism and Christianity. The starting question is how to reconcile the faith with scientific findings. At the top of our knowledge is supernatural revelation, given by God and preserved in the Church. At the bottom we have our human effort to understand nature. This effort—if conducted in a systematic and experimental way—takes the form of science. The guiding principle is that the two books—the Bible and the book of nature—must be compatible. This principle is founded on the belief that both books speak about one reality created by one God. Diagram 22A shows different attempts to reach the science–faith synthesis.

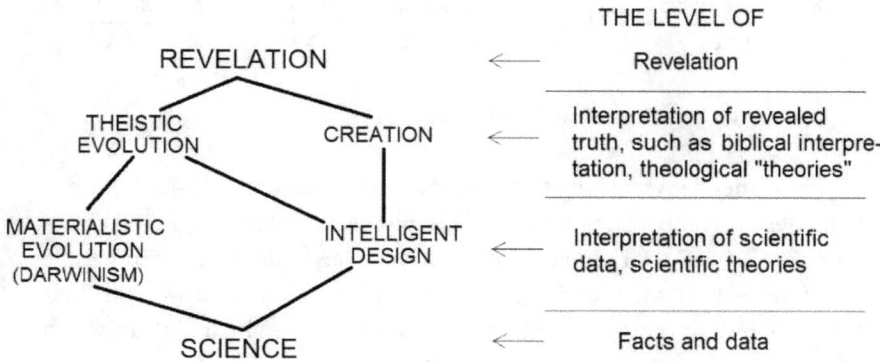

Diagram 22A. Different ways of reconciling science and Christian revelation.

According to the diagram, there are two ways of interpreting biological facts and data: (1) materialistic evolution, which resorts to chance and necessity alone; (2) intelligent design, which in addition to chance and necessity acknowledges intelligent causation. (Keep in mind that materialistic evolution is a broader term than neo-Darwinism. It includes all evolutionary mechanisms that do not involve any power beyond chance and necessity. In the current debate, the most common type of materialistic evolution is neo-Darwinism.)

Revealed truth regarding origins may also have two interpretations: (1) theistic evolution, which is the theological concept maintaining that God used the natural processes of evolution in creating the natural order; (2) belief in creation, which is the theological concept stating that after *creatio ex nihilo* (the first creation of space, matter and energy), God supernaturally formed the natural order during a period of time described in the Bible as the six days. These two interpretations are mutually exclusive, because creation is direct and supernatural, whereas evolution is a natural process working as a secondary cause. Lines in diagram 22A mark the transitions between different concepts. Based on these lines, we can build a table of three models representing three interpretations of the science–faith dialogue present in contemporary Christianity (see Table 5).

| Model A | Model B | Model C |
| --- | --- | --- |
| REVELATION | REVELATION | REVELATION |
| theistic evolution | theistic evolution | progressive creation |
| Darwinian evolution | theory of intelligent design | theory of intelligent design |
| SCIENCE | SCIENCE | SCIENCE |

Table 5. Three models of faith-science synthesis in Christianity.

**Model A:** According to the proponents of this science-faith model, the best current interpretation of the biological facts and the origin of species is (neo-)Darwinian evolution. In order to make this compatible with revelation, the proponents of this model need to interpret the Genesis account as theistic evolution. In this approach (which is also used in Model B), the common stance is that the Book of Genesis only tells us "that" God created the universe, but not "how" he did it. Therefore, theology (the Bible) only tells us that God is the first cause and the ultimate source of all being, while science explains the "mode" of the formation of the universe. With regard to the universe as a whole, its formation is explained by a general theory of evolution, whereas the origin of species in biology is explained by some material evolutionary mechanism, with neo-Darwinism being currently the most accepted. According to Model A, creation out of nothing occurred at the beginning of time, and divine providence works throughout the entire history of the universe. However, it is not clear whether the Big Bang is the effect of the first creation or perhaps marks just one of many beginnings in a multiverse. The proponents of Model A include Kenneth Miller, Rev. Michal Heller, the late Archbishop Jozef Zycinski and George Coyne, SJ, and scholars connected with the BioLogos Foundation.

**Model B:** According to the proponents of this model, the materialistic mechanism of evolution definitely cannot account for the origin of species, yet this does not mean that biological macroevolution did not occur. The proponents of this model usually acknowledge that science currently does not offer any mechanism that can satisfactorily explain the origin of species. However, they also maintain that the general theory of evolution, as long as it is kept within "proper scientific limits," does not contradict any dogma and thus is acceptable for Catholics. Sometimes, they distinguish between the "occurrence" of evolution and its mechanism. Though the mechanism maybe unknown to us, the "occurrence" of evolution is a "fact" that should not be questioned. They are not clear on whether the general theory of evolution (i.e., the evolution of nature) includes biological macroevolution, or evolution is to be understood as merely change over time. The proponents of this model reject the idea that chance and necessity alone can account for the staggering diversity of forms and innovations in nature. They also believe that an intellect of some kind must have taken part in the creation of biological diversity. They do not explain, however, in what way the intellect "tinkered" with nature. Some affirm that "logos" or "reason" understood in a very general sense guides natural phenomena, and others propose evolution "front-loaded" by God, or some form of cooperation between God and nature. They all agree that biological facts are inexplicable without reference to intelligent causation. In this sense, they are proponents of intelligent design, even if they do not explicitly refer to the theory. Hence, in the table we attribute to them the belief in intelligent design as the best interpretation of scientific (primarily biological) facts. At the same time, however, they advocate (either explicitly or implicitly) the evolutionary interpretation of revealed truth. They usually state that "in principle" biological macroevolution does not contradict the Bible, sacred tradition, or Church dogmas regarding creation. They also see a connection between the evolutionary interpretation of the creation story and intelligent causation discovered in nature. It appears that the teachings of John Paul II and Benedict XVI are best summarized by Model B. The positions of

Cardinal Christoph Schönborn, biochemist Michael Behe, and a good number of traditionally minded Catholic philosophers of nature would also fall within this model.

**Model C:** According to the proponents of this model, the best interpretation of scientific facts and data in biology is intelligent design, and the best interpretation of revealed truth is progressive creation—the idea that God acted directly and supernaturally at some points during the time of the formation of the universe. They agree with the proponents of Model A that the universe was created out of nothing at the beginning of time; however, they tend to associate this event with the Big Bang and reject the multiverse hypothesis. They also agree with the proponents of Model A and B that nothing happens outside of God's providence, whether these are chance and unguided or planned and purposeful events. However, the proponents of Model B and C (in contrast with the proponents of Model A) agree that chance and necessity cannot account for the emergence of species. Hence, either implicitly or explicitly, they adopt the theory of intelligent design.

The proponents of Model C differ from the proponents of Model A and B in their understanding of revealed truth. They do not agree that biological macroevolution (universal common ancestry and transformation of species) can be reconciled with Genesis, especially if it is read in accordance with sacred tradition. They maintain that the traditional belief in creation, the same that was taught by the Church until 1950, does not contradict any data of modern science. On the contrary, it is quite compatible with what the actual evidence suggests. They also see harmony between the historical and literal interpretation of Genesis, the principles of classical philosophy, Church pronouncements, and true scientific evidence. At the same time, they do not think that a literal and historical interpretation of Genesis implies that the universe is only several thousand years old. In fact, Genesis itself suggests that the biblical "day" corresponds to a period of time longer than twelve or twenty-four hours. For this reason, the proponents of Model C do not see a contradiction between the scientific evidence and the biblical message related to the natural history of the universe. They acknowledge that the age of the universe is a legitimate scientific question and that science alone is competent to determine the answer. Among the proponents of this model are retired auxiliary Bishop of Salzburg Andreas Laun, theologians such as Robert Stackpole, Catholic scientists such as Ann Gauger, and Protestant scientist Hugh Ross and some individuals connected with his organization Reasons to Believe.

## c. Toward the New Science-Faith Synthesis

None of the three science-faith models is currently favored by the Church. In fact, the Church authorizes—sometimes implicitly, sometimes explicitly—mutually exclusive ideas about origins. Some Catholic scholars believe that the issue is already settled. They come to this conclusion, however, without taking into account the real state of the debate, and without including all of the different positions advocated by different Catholics. Another group believes that the issue cannot be addressed by the Church in greater detail, because these questions belong to science, not to the Church's magisterium. Still another group advocates a return to the traditional "creationist" positions, without any regard to the achievements of modern science. In our opinion, however, the progress of science, along with a renewal of Catholic doctrine on creation, may generate a new science-faith synthesis that was unavailable to previous generations. Moreover, we disagree with those who think that the issue cannot be definitively resolved by the Church.

The Catholic Church used to have a specified doctrine on creation, but over the last century it was largely abandoned under pressure from the so-called "scientific community." There are many reasons for the withdrawal of the Church from teaching the traditional doctrine on creation. Some of them we explained in the first chapter of this book. However, the very fact that the Church once held a clear position regarding at least the origin of man suggests that some of her competence has been given up too hastily. Science makes progress in the understanding of nature, but it cannot remove the limits of science and enter those areas that are explainable by theology alone. The Church should reclaim her competence and authority to present authoritative answers to the questions of origins. This is even more desirable if we realize that today's science has opened the door to a much better understanding of Genesis than what was available to any ancient, medieval, or early modern theologian. Believers should not be surprised that God made the truth about origins available only little by little, as human knowledge progressed and expanded. After all, if both books—nature and the Bible—were written by God, both must be read to the end in order to make known the full message they contain.

In our opinion, the time has come to make a new coherent picture out of all available data—theological, philosophical, and scientific.

In our presentation of the three models A, B, and C (in the previous subsection), we avoided judging any of them. It is obvious, however, that they are not equally probable or justified. Now we should comment on each of them in light of all the facts presented in the previous chapters. This will help us to establish which one of the models conveys the best science-faith synthesis in light of current human knowledge as discovered by science, philosophy, and theology.

**Model A** accepts the (neo-)Darwinian mechanism of evolution. However, as we explained above (in II,4), this mechanism cannot account for the emergence of a single gene, let alone the dramatic increase in genetic information that was needed in transitions between species and, to a far greater degree, in events such as the Cambrian explosion of life. Experiments repeatedly show the very limited capability of random mutations and natural selection to produce biological novelty and the very high requirements that biological designs have to meet in order to make life possible.[53] Moreover, neo-Darwinism does not explain (does not even address) the problem of epigenetic information, which is indispensable for organisms to develop from the embryonic state. For these and other reasons, the neo-Darwinian theory of the origin of species has reached a dead end and cannot be proposed as the best interpretation of biological facts and data.

Model A assumes that theistic evolution is compatible with the neo-Darwinian mechanism of evolution. But this mechanism implies that everything in biology began through pure chance (i.e., random mutations), whereas theistic evolution states that God is somehow present in the process. The proponents of Model A cannot explain how the evolutionary process can be completely random and divinely guided at the same time. Hence, contrary to what Model A affirms, theistic evolution is incompatible with neo-Darwinism.

Model A has in common with Model B the evolutionary interpretation of the book of Genesis. Unfortunately, the popular saying that there is no contradiction between biological macroevolution and Genesis is rarely confronted with the actual problems arising whenever we ask about any details. Surely, theistic evolution agrees with a number of

---

53. For evidence, see, e.g., books by Michael Behe (*Darwin's Black Box, The Edge of Evolution, Darwin Devolves*) and Stephen Meyer (*Signature in the Cell, Darwin's Doubt*).

very general statements that are implied in Genesis, such that God loves all creation, that everything ultimately comes from God, or that everything is planned and desired by the Supreme Being. But as soon as we ask slightly more specific questions, insurmountable problems emerge. For instance, Genesis explicitly and repeatedly teaches that God created animals and plants according to their kinds. How can this be reconciled with the idea of universal common descent? Genesis teaches that man's body was formed from *the dust of the earth*, which clearly indicates that inanimate matter was used to produce the human body by direct action on God's part. How can this be reconciled with the idea of preexisting *living matter* serving as a substrate for the production of the first human being? Genesis clearly teaches that in creation God acted in a particular way for a definite time and that this action ceased when God rested on the seventh day. Evolution, however, is a natural process that operates according to the laws of nature. As such, it can hardly be called a particular divine action; it never stops and never fundamentally changes. It continues as long as natural laws govern nature. How, therefore, can theistic evolution explain the fact that creation was finished once for all with the creation of man? Finally, theistic evolution cannot be reconciled with all the interpretations and clear statements on creation by Church Fathers, Doctors of the Church, popes, saints, and the majority of theologians throughout nineteen centuries of Church history, both long before and several decades after Darwin.

The alleged compatibility between evolution and Genesis is not a conclusion derived from any serious reading of the text, let alone any serious theological study. It is imposed on the text in order to make it compatible with one poorly documented theory of nature. The alleged compatibility is possible only if Genesis is read without connection to sacred tradition and after abandonment of the historical and literal sense; neither of these is a Catholic approach to biblical texts. Thus, Model A fails in its understanding of the supernatural revelation expressed in Genesis as well as in its understanding of scientific data. As such, this model cannot be a valid science-faith synthesis for the Church.

**Model B** is similar to Model C in its interpretation of the scientific data. Both models accept intelligent design as an explanation for some biological structures or events. As we noticed, random mutations and natural selection cannot account for the increase in biological novelty observed in the fossil record. However, the increase in information in the history of life can be explained by referring to intelligent causation—the only known cause capable of producing new functional information. Hence, intelligent design is a better interpretation of biology than neo-Darwinian or materialistic evolution. Model B assumes that there is compatibility between theistic evolution and intelligent design. And this is true, because new information required to build new forms of life could have been introduced into biology through the mediation of some evolutionary mechanism. At least this option is not excluded by the theory of intelligent design.

ID, in principle, does not exclude universal common ancestry or transformation of species. However, these two ideas cannot be reconciled with classical metaphysics and the literal historical reading of Genesis. Therefore, even though Model B meets the expectations of science, it can hardly be considered compatible with sound philosophy and the original Christian theology of creation. In order to maintain biological macroevolution, Model B dismisses Genesis as understood in the Catholic interpretative tradition. For these and some other reasons,[54] this science-faith model is also unsatisfactory.

Only **Model C** meets all of the requirements coming simultaneously from science, philosophy, and faith. It adopts intelligent design, which is a better (and adequate) explanation for the origin of the vast quantity of useful and coherent information that we find in biology. Progressive creation is a theological interpretation of revealed truth regarding the origins of the universe. Since it adopts the supernatural formation of the universe, it is compatible with the biblical message as well as sacred tradition. It also makes room for the historical and literal reading of Genesis, which is the traditional position of Fathers and Doctors of the Church. Moreover, because progressive creation does not entail a young universe, it is also compatible with modern scientific evidence regarding the age of the universe. Additionally, progressive creation is entirely compatible with the theory of intelligent design, because design found in nature could have been introduced into biology through direct divine causality. Model C, thanks to its rejection of universal common ancestry and the transformation of species, is also compatible with the highest standards of classical metaphysics, as well as the teachings of the greatest theologians,

---

54. Other reasons are listed in our book *Aquinas and Evolution*, in the last section (Excursus 2). I presented there some scientific reasons why at least the "front-loaded" type of evolution is impossible. Therefore, an opinion that theistic evolution, which assumes that God introduced design by an evolutionary process, is reconcilable with science is not tenable. See *Aquinas and Evolution*, 237–54.

such as St. Augustine and St. Thomas Aquinas. Model C is therefore the best candidate for becoming a new Christian science-faith synthesis.

Diagram 22B summarizes our analysis of the three models. Some of the transition lines have been marked with an X. This means that these transitions, even if readily adopted by some scholars, are actually impossible because the ideas they connect are incompatible.

Model A fails at the transition between scientific facts and scientific theory, between theory and the theological concept, and between the theological concept and revelation (theistic evolution–revelation). Model B fails at the transition between the theological concept and revelation. Only Model C synthesizes all levels of the debate, without generating difficulties in any of them.

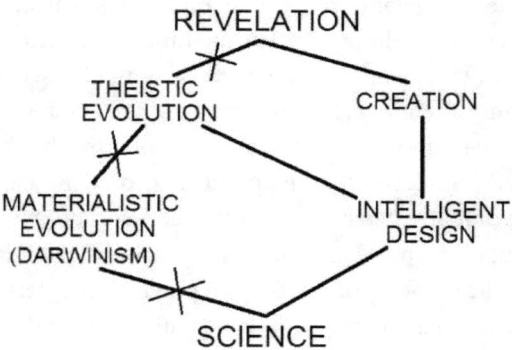

Diagram 22B. The actual way of reconciling science and Christian revelation.

### d. The Renewal of the Catholic Theology of Creation

Today, the Church faces a number of new challenges, both from global culture and from dissidents within itself. John Paul II called this a war between two cultures—the culture of life and the culture of death. It appears that the battle fought by the Church and all people of good will is strongly focused on man—his identity, his place and role in the world. To understand the true nature of man, we need to establish who he is among other beings in the universe. This, however, entails also the questions regarding the ultimate destination of man as well as the questions about his origin (when and how he began to exist). Only after we find the correct answer to these questions can we secure the proper place for man in the universe.

Current theology, as well as the Church as a community, is being renewed in many aspects. This has been happening as a result of cultural challenges from the world and an attempt by believers to find solutions to the moral problems of global civilization. Thanks to Pope John Paul II, the Church discovered the personalist approach, which greatly helps in adequately addressing the aberrations of the culture of death. In addition, classic philosophy is finding new outlets and hubs at newly founded or renewed Catholic universities. Many deep interpretations and insights are being developed in disciplines such as natural law, moral theology, bioethics, family studies, and the theology of the body. We also see a decline of extreme biblical criticism and an overall focus of the institutional Church on religious problems, rather than on the political allegiances that burdened her so much in the eighteenth and nineteenth centuries. Despite the signs of crisis, everywhere in the world we can also see obvious and encouraging signs of renewal. The springtime of the Church is at hand.

There is, however, one dimension of Catholicism that has been declining since early modernity, to the point that today we can speak of a huge crisis that urgently needs to be addressed. Not too many Catholics realize the extent of the problem, and this is precisely why it is so serious. This is the Catholic theology of creation. In previous chapters we saw that the traditional Christian interpretation of creation was challenged in modernity by new scientific discoveries. Since the Copernican revolution, people of faith have been struggling with the question of how to save a dogma when science seemingly contradicts it. After Darwin, the conflict dramatically accelerated and entered a new phase. From then on, science was used not just to explain the operation of the universe in its current state (the competence widely recognized by Christianity even before the scientific revolution), but also to address questions of origins, such as the origin of man and species. In the twentieth century, after decades of struggles, a majority of Christians gave in to theistic evolution which is poor science and even worse theology.

For ancient and medieval theologians, the claim that any being could participate in creation as an active help to God was serious heresy. Yet today, this approach has become almost a universal solution among the believers of different Christian denominations. Theistic evolution seems to neatly combine science and faith and peacefully resolve the long-standing conflict. It is proposed as an easy solution to a difficult problem. Yet easy solutions are not always correct. When applied to faith, they should generally be received with suspicion, especially when the solution mitigates a conflict between Christianity and a materialistic worldview. Theistic evolution is not an exception here, and this is why, instead of being unreservedly accepted and promoted, it should be scrutinized with special attention.

If we look at theistic evolution from a perspective stripped of popular and simplistic interpretations in biology, philosophy, and theology, we immediately notice the shortcomings of its various assumptions. Whenever we ask about the details, theistic evolution reveals its inadequacy. It begins with an importation of universal common ancestry and transformation of species into Catholic theology of creation. This invasion of unfounded philosophical claims into theology creates a caricature of the Christian doctrine which is presented as a quasi-official Church position.

The foundational principle behind theistic evolution is that the Bible does not tell us anything about *how* God created; it only says *that* God created. Theistic evolution, while claiming to be a science-faith synthesis, actually disregards the Bible on one hand and strongly promotes scientific theories over scientific data on the other. In effect, theistic evolutionists either completely abandon or dramatically downplay the role of creation. It is not unusual for them to say that belief in creation is not nearly as important as belief in the Incarnation or the Resurrection of Christ. They thereby "spiritualize" Christianity in a peculiar way, by putting overwhelming emphasis on the history of salvation to the detriment of the history of creation. On one hand, theistic evolutionists allow the materialism of evolutionary theories, and, on the other, they distance Christianity from natural history. For some theistic evolutionists, the overall role of the doctrine of creation is so insignificant that they willingly accept any theory, conjecture, or mere assumption, as long as they enjoy the support of the so-called "scientific community," no matter how blatantly they would contradict the Faith. In this sense, theistic evolution is not a true synthesis. It is, in fact, a disconnection between faith and science, where science is authorized to explain everything visible and material, and faith is restricted to what is spiritual, eschatological, and invisible. Theistic evolution, rather than being a Christian answer to the challenges of the present age, is actually a vague combination of materialism and fideism.

We agree with theistic evolutionists that in some sense the Gospel, the good news of salvation in Christ, is more important than the origin of the universe and the "mode" of divine creation. Yet, at the same time, we need to stress two things. First, the teaching on creation is not accidental to the Faith; it belongs to the essential doctrines, and it differs substantially from any ancient or current mythologies (see III,1,a). Second, the teaching on creation influences our understanding of other truths of faith, including the Incarnation and the Resurrection, even if the connection is not immediately obvious.

Classic theology speaks about the *nexus mysteriorum*—the net of mysteries. This refers to the idea that all truths of faith are connected and dependent on one another. Hence, the removal or distortion of one of them necessarily leads to the destruction of others. For these and other reasons, the Church of today faces a great need for a renewal of the Catholic theology of creation. This does not mean a mere return to teaching the traditional writings "on God the Creator" (*de Deo creante*) as if nothing has happened in science and culture over the past few centuries. One extreme cannot be replaced with another. The gratuitous embracing of bad science in theistic evolution cannot be countered with an equally gratuitous rejection of well-established scientific data in creationism. Hence, the renewal of the Catholic theology of creation entails a whole new science-faith synthesis and the reimplementation of the biblical paradigm in our understanding of the origins of the universe. The renewal must be based on sound Catholic doctrine enriched by new scientific discoveries, such as intelligent design in biology, the theory of fine-tuning in physics, the Big Bang in cosmology, and deep time in geology. The new science-faith synthesis should be based on the best scientific data rather than outdated theories, realistic philosophy (*sana philosophia*) rather than imaginative speculations, and a serious approach to the book of Genesis, interpreted (in accordance with the entire Catholic tradition) as historical truth rather than metaphor or literary fiction.

### e. The Benefits of the Renewal

What are the greatest challenges for the Church in our times? Few would deny that these are the protection of human life and the defense of a proper understanding of human dignity, marriage, family, and the sexual identity of the human person. This situation reminds us of the Gospel passage in which the Pharisees come to Jesus to ask about divorce. In his response, Jesus appeals to Genesis. He says that from the beginning it was decided that man and woman were created to be together, and once joined in the marital vow, they should never be separated. We see that Jesus appeals to Genesis to teach people about the nature of marriage and human love. For centuries, the Church followed the way indicated by the Master—whenever a controversy regarding marriage arose, popes and theologians appealed to Genesis. And this form of argumentation is entirely justified, because Genesis recounts the true origin of humanity, regardless of whether people believe it or not.

Once the literal and historical reading of Genesis is abandoned, some of the greatest arguments of Catholic apologetics are lost. Indeed, without the biblical account, there is no way to understand human nature and ultimately no way to justify human dignity. In contrast, belief in creation provides a strong argument for the dignity and special destination of every human, with respect to both the body and the soul. The direct creation of the first human body tells us that it was designed from scratch to accomplish the objectives of the soul. Thus, man is called to supernaturality in both aspects of his nature—the corporeal and the spiritual. Moreover, this means that his sex—his identity as a man or a woman—was also designed and realized by God from the very beginning. Masculinity and femininity are two forms of humanity, two ways in which the human design has been actualized.

In contrast, belief in the evolutionary origin of the human body raises doubts about the permanent character of human sexual identity. If sex was created by essentially random evolutionary processes, then why shouldn't man, endowed with reason and will, modify it according to his own whims? In the evolutionary perspective, how can we know that sex is not just a cultural concept, the effect of social adaptation that may be beneficial one day and detrimental another? If so, why wouldn't humans tinker with their bodies, trying to change their sex? If humans descended from apes through many mysterious humanlike creatures, we do not even know when humanity begins or whether humans constitute one species.

The founders of the modern theories of race, later adopted by the Nazis in their extermination of other nations, were all explicit Darwinists.[55] Ernst Haeckel, for example, believed that the distance between the

---

55. Richard Weikart, a historian, provides detailed historical evidence for the intellectual connections between Darwinian ideas (i.e., the struggle for life and the survival of the fittest) and modern racism and social Darwinism. See his *From Darwin*

highest and the lowest of the human races is greater than the distance between the lowest humans and the highest apes. If the evolutionary origin of man is accepted, then the inequality of men, racism, and eugenics follow as almost necessary logical conclusions. Regardless of whether or not one feels personally offended by the idea of man descending from lower animal, this idea strongly supports a number of concepts and policies that are directly opposed to human dignity. In order to make the argument for human dignity coherent and convincing, believers need to follow Christ and return to the beginning, to Genesis, to the full truth about the creation of man.

Let's imagine that the Church openly adopts a new science-faith synthesis that includes the original teaching on creation (i.e., what was taught until the 1950s) without rejecting any modern scientific data. What would Catholic scholarship and apologetics look like? What would the possible benefits be?

Initially, there would be a great outcry about how faith in general and Catholicism in particular destroys science and act contrary to reason. But after this initial frenzy, the emotions would be replaced by a deeper reflection, and ultimately most Catholics would realize that the new science-faith synthesis not only promoted the true faith but was also beneficial for science itself. It would strip science of the unfounded and reductive theories that distort data and hinder correct, rational conclusions. Moreover, a number of Christians, among them many scientists, currently struggle whenever they learn about the neo-Darwinian story in the "science literature" on one hand and about creation in the Bible on the other. The new synthesis would help them to be consistently faithful and orthodox believers while pursuing science according to the highest academic standards.

In philosophy, the new synthesis introduces excellent opportunities for the renewal of realistic reflection on nature and particularly the revival of classical metaphysics. Once Catholic philosophers realize how the principles of biological macroevolution contradict the principles of sound philosophy, they will employ philosophy to defend the essential truths of the Faith, as was done for centuries before Darwin. Moreover, after they realize how far-fetched the evolutionary claims are, even in the context of science itself, they will see anew how metaphysics is not just a product of the ancient minds, but an authentic insight into the very

---

*to Hitler*. In another book, Weikart shows how evolutionary theory influenced Hitler's views; see his *Hitler's Religion*.

nature of reality—insight that is as true for us as it was for Aristotle or Aquinas. In this way, perennial philosophy will regain its authority in Church teaching and culture, and also the proper role and limits of scientific and philosophical disciplines will be established.

Theology will undergo a great renewal. Since the modernist crisis of the early twentieth century, biblical scholarship has been detached from dogma and sacred tradition, and focused on reductionist and rationalistic interpretations that have often been used against the teachings of the Church. The renewal of the historical interpretation of Genesis will reconnect the Bible to the interpretative and doctrinal traditions of the Church Fathers and the Doctors of the Church. It will also reinstitute the Bible as an authoritative source of faith, in contrast to what it is now, namely, an object of merely a human intellectual endeavor.

The renewed teaching on creation will also provide the deepest foundation for the theology of the body. If the first human body was made from scratch directly by God, it follows that each human body produced in the natural course of generation carries the same original design. This design serves primarily the spiritual needs and goals of man, and this is why the body is holy and called to holiness. These are just a few glimpses into how Catholic scholarship will benefit from the renewed teaching on creation.

## 6. Christianity and Culture

### a. Richard Niebuhr's "Christ and Culture"

During my basic studies in theology, one of my assignments was to read the book *Christ and Culture* (1951) by the Protestant author Richard Niebuhr.[56] In the foreword to the fiftieth anniversary edition, one can read that Niebuhr's book is among just two or three that have been considered classics in theology throughout the twentieth century, and that its appeal is not likely to have been exhausted when 2051 rolls around. No wonder—*Christ and Culture* attracts readers with an easy style and impresses with its breadth of view, inclusiveness, and erudition. The objective of the book is to present different types of believers according to how they relate to culture. The author elaborates upon five main models: (1) Christ against culture; (2) Christ of culture; (3) Christ above culture; (4) Christ and culture in paradox; and (5) Christ the transformer of culture.

---

56. H. Richard Niebuhr, *Christ and Culture*.

A typical representative of the first model (Christ against culture) is Tertullian, and in modern times Leo Tolstoy. The radicalism and exclusivism of this model raises, however, a few theological problems, such as how to reconcile reason and revelation, or law and grace. There is an emphasis on the sinfulness of culture and the need to uproot its vices. According to Niebuhr, Christians of the first model have problems with a coherent view of two realities—Christ as the absolute ruler of the universe on the one hand, and the Spirit immanent in creation and the human community on the other. Their mistrust directed toward nature brings them to spiritualism and rejection of culture or even of the historical Jesus. This is why, Niebuhr concludes, "radical Christianity, important as one movement in the church, cannot itself exist without the counterweight of other types of Christianity."

As if on the opposite end of the spectrum, there is the Christ of culture—a model represented by people who nearly identify Christian belief with the predominant culture of their times: "They feel no great tension between church and the world, the social laws and the Gospel, the workings of divine grace and human effort." Peter Abelard, a medieval scholar, and the mainstream Protestant philosophers are singled out as examples of this approach. Locke, Leibniz, Kant, Thomas Jefferson—they all consider Christ the great teacher, the one who leads people through their culture toward moral perfection. Later Schleiermacher, Hegel, Emerson, and Ritschl depart from "religion within the limits of reason" to adopt the "religion of humanity." In these philosophies, the "acculturation" of Christ reaches its peak, and ultimately Christianity turns into a merely cultural phenomenon. Niebuhr summarizes: "It becomes more or less clear that it is not possible honestly to confess that Jesus is the Christ of culture unless one can confess much more than this."

When speaking about Christ above culture, Niebuhr mentions the notion of the "church of the center," which he considers to be mainstream Christianity, or the Christianity of the majority. Those who represent this model see culture as founded on nature, which, in turn, is rightly ordered by the Creator toward Christ as its goal. There is also a high degree of agreement between different representatives of this model regarding the universality and radical nature of sin. While the radicals tend to exclude Christianity from the devastating influence of sin (the holy amidst the sinful world), the cultural Christians tend to deny that sin reaches into the depths of human personality. Christians of the center, in turn, recognize the reality of sin, the need for redemption and the aid of the sacraments.

Niebuhr tells us that regardless of particular differences, Aquinas and Luther are still closer to each other on the subject of grace than either of them is to the Gnostics or the modernists.

Finally, the author introduces the idea of a "synthesis," which is the main characteristic of the church of the center. The synthesists are those who see connection and interdependence, rather than isolation or confusion, between human deeds and divine grace, law and charisms, the natural and the supernatural, human freedom and God's omnipotence, etc. Their approach is succinctly expressed in "both/and" rather than "either/or" formulas. The "synthetic" answer to the problem of Christ and culture was given, for example, by Clement of Alexandria, Thomas Aquinas (who, in Niebuhr's words, "is probably the greatest of all the synthesists in Christian history"), and the Anglican bishop Joseph Butler.

The two remaining models (Christ and culture in paradox and Christ the transformer of culture) are in their own ways middle grounds between "Christ against culture" and "Christ of culture." However, they also differ from the synthetic type just described.

"Christ and culture in paradox" may be best described as a dualist model, though different from the Manichean dualism of good and evil kingdoms of God and Satan. It is more about the dualism between me, a believer, and God the Creator, the demands of earthly life and the demands of the Gospel. Even if men accept the revelation of Christ and divine grace they still remain outside of and far away from him. Grace belongs to God, whereas sin is in man. Human nature, as well as the totality of human culture and activity, is corrupted by sin. "Thus in the dualist's view the whole edifice of culture is cracked and madly askew," says Niebuhr.[57] In this view, the Gospel with its demands must appear as a paradox, a demand of something overwhelmingly difficult, possible only by grace. In law and grace, God's wrath and mercy remain in opposition. Among the proponents of this approach Niebuhr mentions St. Paul, the second-century gnostic Marcion, Martin Luther, and the Danish existentialist Soren Kierkegaard. The contemporary dualists consider religion and science two realms that are neither in conflict nor in any positive cooperation. Roger Williams presented a dualistic theory of Church and state, whereas Nikolai Hartmann accepted a permanent antithesis between Christian faith and cultural ethics.[58]

---

57. H. Richard Niebuhr, *Christ and Culture*, 155.
58. H. Richard Niebuhr, *Christ and Culture*, 183–84.

The last model, "Christ the transformer of culture," considers culture neither the domain of evil nor the realm already completely cured from sin; culture demands conversion. The proponents are similar to dualists and synthesists in their understanding of Christ more as Redeemer than as the giver of a new Law. Jesus is not the one who externally judges human behavior, but the one who tries human hearts and intrinsically accompanies people on their way. Unlike dualists, however, conversionists represent a more positive attitude to the created order. Redemption does not surpass creation; rather, creation is the basis for the salvific actions of God. Hence, the Word (*Logos*) participates in creation throughout time, and not just momentarily in the event of Jesus of Nazareth. As a result, sin is not something that ruined nature, but something externally imposed that may be overturned by conversion. For the conversionists, says Niebuhr, "Culture is all corrupted order rather than order for corruption, as it is for dualists. It is perverted good, not evil; or it is evil as perversion, and not as badness of things."[59] As the best example of conversionists' convictions, Niebuhr presents St. John's teaching in the Fourth Gospel. Another great representative of this approach is Augustine. In modern times, a great example is the English theologian Frederick D. Maurice.

This is, in short, the content of the book *Christ and Culture*. The author successfully avoids many oversimplifications that could easily sneak into such a comprehensive view of Christianity and culture. It is impressive how Niebuhr captures permanent Christian tendencies throughout history.

Nevertheless, there is a problem—hardly noticeable, yet quite dangerous—in the way he presents the matter: Niebuhr evaluates and categorizes the figures of the past in light of culture. He considers culture a stable phenomenon, a kind of constant reality, to which different individuals relate differently according to how they understand Christ. The way they express Christ in their lives and teachings determines their approach to culture.

To explain the problem, we will reduce Niebuhr's typology to just three positions and then present it in diagrammatic form. The three positions are (a) Christ against culture, (b) Christ of culture, and (c) Christ and culture. Diagrams 23A and 23B represent Niebuhr's thought.

---

59. H. Richard Niebuhr, *Christ and Culture*, 194.

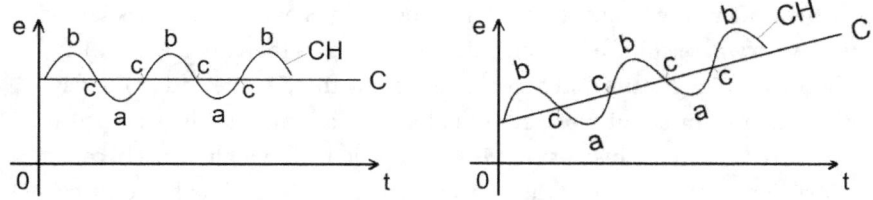

Diagram 23A. Christianity vs. culture over time, three positions.
Diagram 23B. Christianity vs. culture over time, including the progress of culture.

The o→t axis represents time from the Incarnation to now. The o→e axis represents the demands of culture regarding religion. The straight line **C** in Diagram 23A is horizontal, representing the constant expectations on the part of culture, as Niebuhr's book implies. The author does not mention any progress in culture, but neither does he exclude it. Hence, if we assume that culture as such also progresses over time, its expectations would be higher in subsequent epochs. For instance, our scientific era, with all its cultural, technological, and academic achievements, would require a more rational religion, i.e., a religion justified by reason and capable of critical self-reflection. Unlike primitive peoples, contemporary societies are not satisfied with a revelation that flatly contradicts scientific facts or common historical experience. Similarly, our democratic era would reject a totalitarian religion, or a conflation of religious and secular authority. This is why, in Diagram 23B, the cultural constant **C** increases over time. Culture must be understood according to Niebuhr's definition as something morally neutral and generic. Hence, the progress of culture might be imagined as an expansion from a middle point toward the edges of a circle that represents all human activity. In this imagery, even though culture progresses, it does not become better or worse. Except for this one factor of cultural progress, the two diagrams (23A and 23B) are identical.

The curve **CH** represents the oscillation of Christ (for Niebuhr, Christ means the way in which individuals adopt and express their Christian faith) around the cultural constant **C**. Points **a** and **b** represent extreme positions: **a** stays far below **C**, because it stands for the model "Christ against culture." On this view, culture is for Christianity a sordid enterprise that should be avoided and diminished. The peaks labeled **b** stand for the model "Christ of culture." On this view, Christianity is open to compromising its own identity for the sake of synthesis with culture. "Christ of culture" wants to meet all expectations of culture. Christianity is always where culture is, even before culture gets there. In a way, Christianity mimics culture and finally identifies with it. At the end of the day, there is no Christianity but culture alone, which moves Christian spirit wherever it wants. This position is symbolized by the high peaks because Christianity even exceeds the expectations of culture. Finally, point **c** is where Christianity exactly meets the expectations of culture—it neither goes too far nor stays behind. Synthesists are those who meet culture at the exact place of its expectations.

The oscillation of curve **CH** is symbolic. It does not mean that Christianity oscillates around culture periodically according to established intervals. In fact, Niebuhr shows that the types of Christianity represented by points **a, b,** and **c** may be present simultaneously in all eras from the Incarnation until now. The reason for the sinusoid **CH** is to highlight one crucial idea pervading all of Niebuhr's book: Christianity is not constant, culture is. Christianity is evaluated in the light of culture, not culture in the light of Christianity.

### b. Can Christianity be Countercultural?

No doubt religion can be distinguished from culture—these two realities are not reducible to one. However, Niebuhr implies something more, namely, that culture may be neutral in terms of religion and essentially constant in its relation with religion. In the Niebuhrian sense, Christianity is what Christ brings to the world—his expectations and demands, his truth and morality, his precepts and grace. But isn't it true that what Christ brings to the world is more stable and objective than what humanity produces on its own?

Here is the problem. Niebuhr implies that culture is a permanent human creation that constitutes a reference point for the reality of Christ. Niebuhr swaps the order of priority between the two realms (Christianity and culture) and makes culture the criterion according to which Christianity is evaluated. Culture becomes normative in relation to religion, and not vice versa.

It seems more adequate, however, to see Christianity (Christ's revelation) as something objective and unchangeable, whereas culture is something that fluctuates over time, specifically when it interacts with externally revealed religion. The problem with the perspective adopted by Niebuhr is that it is not possible to define cultural demands properly without reference to religion. Whether we consider morality, faith (the question of truth), or a worldview, in each case supernatural religion provides a permanent reference point for culture. Whereas culture might be misled by the lower desires of mankind or other natural factors, true religion has an inherent objectivity established by God. Niebuhr's approach requires an essential reformulation. In Diagram 24 evolving culture is set up in reference to the permanent reality of Christ.

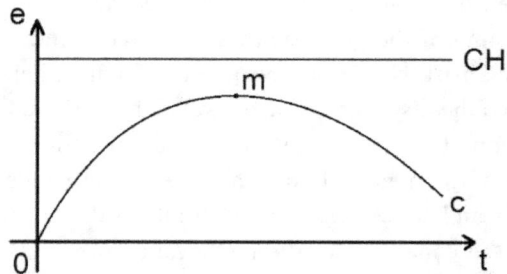

Diagram 24. Christianity vs. culture: the actual historical development.

As in Diagrams 23A and 23B, here also axis o→e represents the level of expectations. This time, however, Christianity is a constant that represents the expectations (line **CH**) posed by faith to human progress. We can say that Christ in his proclamation of the Gospel outlined the shape of the highest possible development that human culture could ever achieve. He called humans to create a culture that would meet the ideal set by him. There is nothing corresponding to a cultural constant, because culture evolves over time, specifically when it encounters powerful spiritual inspirations, such as Christianity.

When Christianity entered the world (point o), the expectations of Christ went far above the abilities of culture (i.e., the distance between o and e was greatest). Indeed, early Christians needed to speak out against multiple cultural "standards" that had to be overturned with the help of religion. These included polytheism, caste systems, inequality among humans, slavery, infanticide, divorce, homosexuality, and polygamy, to name a few. The process of transformation of culture, however slow, took place in Europe over the following centuries. After a thousand years or so, culture became Christendom (Lat. *Christianitas*)—one religious and political order in which Christ was officially the norm for all human activity. It is no wonder that Niebuhr found the best example of the synthetic approach among the medieval scholars. It was in medieval Europe that Christ and culture became one, yet not owing to the dissolution of Christ in culture but rather because of the transformation of culture according to Christ. Point **m** in the diagram represents this medieval synthesis of Christ and culture. However, culture can never fully meet the expectations of Christ. There always remains an unattainable horizon that can inspire new generations to be ever stronger and deeper in their transformation by grace (this is why parabola **c** never reaches constant **CH**).

After the time of the great synthesis, human culture again alienated itself from Christ. Numerous events in different spheres of human activity provoked the dispersion of medieval Christendom. Of particular significance among them was the Protestant Reformation, which led to the destruction of the medieval political order. Scientific developments, such as Copernican heliocentrism and Columbus's discovery of the New World, ruined the coherence of the medieval worldview. The progress of the experimental sciences led to their emancipation from the Church and their isolation from philosophy of nature. Rationalism and idealism created a new realm of philosophy outside of the auspices of theology.

In late modernity the disparity between mainstream culture and Christ becomes greater and greater, approaching the initial state of the Christian era. And not only does culture become an independent realm—it also generates phenomena inherently hostile to the Christian religion, such as the great atheistic ideologies of the twentieth century. This explains why Niebuhr cannot find a good example of the synthetic approach among his own contemporaries: culture is too far from Christ to make the synthetic view obvious or even possible.

In contrast with Diagram 23, in Diagram 24 it is culture that oscillates toward and away from the unchangeable demands of Christ. When the distance between curve **c** and line **CH** is greater, Christianity seems more countercultural because then human culture needs more radical reform to meet the expectations of Christ. Niebuhr's understanding of this relation implies that Christianity is the one that works against culture, whereas in our opinion it is more proper to say that it is culture that may act against Christianity. Culture by itself (i.e., completely isolated from the supernatural order) can never fully understand the human being, his nature, deepest aspirations, and happiness. Consequently, culture without inspiration from religion focuses on immanent ends that are defined differently by different people. This leads to erosion of the universal human values, and ultimately culture ends up in the old pagan errors such as extreme inequalities, tyranny, and diminishing human dignity. Culture, by itself, cannot constitute a subject to which religion might refer.

Our reversal of the reference point from culture to Christ creates a different understanding of counterculturality. Niebuhr's approach implies that being in opposition to culture is something negative that should be avoided. As a result, if we think along Niebuhr's lines, we consider a Christian someone who should look for a synthetic approach to culture. In our understanding, counterculturality is not a matter of free choice. It may be an unwilled, though unavoidable and even necessary, approach to culture, if one wants to remain faithful to Christ.

We can transform Diagram 24 by deriving the level of counterculturality and presenting it as a new diagram (Diagram 25). Here we see how the expected level of tension between culture (society, customs) and the highest demands of Christ changed throughout the two millennia of our era. The level of counterculturality is represented by curve cc.

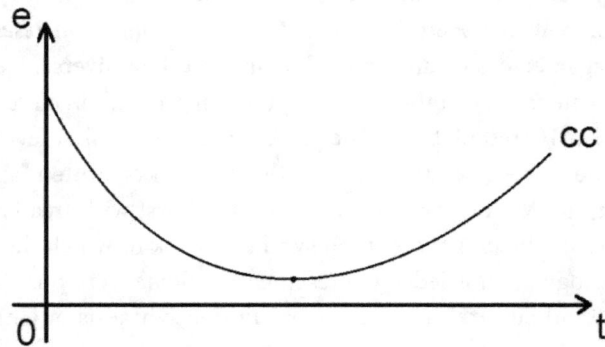

**Diagram 25. Counterculturality of Christianity in its development over time.**

Now we are ready to answer the question: Can Christianity be countercultural, or should it always look for a synthesis with mainstream culture?

First, we should notice that this choice is not a matter of a defined norm; rather, it is a matter of prudence. It is never a true goal of Christianity to challenge culture as such. Christianity has expectations regarding two general spheres—truth about God and the universe (articles of faith) and truth about morality (moral teachings). Christianity challenges culture in these two spheres, and in other spheres only as much as they refer to these two. This is why Christianity can be enculturated in different times and places without compromising its highest expectations. Enculturation occurs when the high expectations of Christ are met in a given culture and the religiously irrelevant spheres of culture remain unchanged.

Let's see how our model describes actual history. In the beginning of Christianity, we see a small group of disciples who go out and call for a total conversion of all humanity. They are like aliens entering society, trying to convert and renew it from the bottom up. Over time, Christianity gains intellectual support in philosophy, which is employed to justify and clarify the faith, but it is difficult to uproot some of the pagan customs. Even though Niebuhr identifies St. Clement of Alexandria as a synthesist, the vast majority of Christian writing during this time was polemical, directed against fundamental pagan beliefs and paradigms. The initial intellectual movement in Christianity is marked by works whose titles begin with the word *contra*. They are clearly in opposition to the dominant views. The greatest theologian of the pre-philosophical Christian era—St. Irenaeus—would make a good candidate for the synthesist. Yet, his main work, *Against Heresies* (Lat. *Adversus* or *Contra Haereses*), is a catalog of pagan errors regarding God, man, and the universe. Irenaeus does not hesitate to correct the misunderstandings of faith in culture. His attitude is quite representative of the general tenor of the first few centuries. When we move on in time, we see that the works written "against" disappear. In the Middle Ages, they are replaced first with treatises and next with the *summae*. A *summa* is a synthetic work in which the entire body of theological knowledge is presented. *Summae* were possible because Christ and culture came very close in one synthesis of Christendom. The requirements of Christianity were not adjusted to make the *summae* possible; rather, it was culture that was transformed by Christ.

In modernity, we see that the roads of the intellect and faith diverge. Christianity undergoes rationalization and becomes more like a conceptual framework carried by culture, rather than a living faith. *Summae* are first replaced by commentaries on the *summae* and later by manuals. But what is a manual if not an abstract thinking about faith that may have no connection with life? We can thus speak of progressive over-intellectualization of faith in the West. The intellect, still somehow inspired by faith, starts generating its own ideas about the universe and God. Even if the authors of modernity considered themselves Christians, they really lost connection with the highest Source. And this explains why Niebuhr finds the most extreme "Christ of culture" type of Christians among the mainstream philosophers of modernity. For Niebuhr, it is Christ who is dissolved in culture, whereas according to our perspective it is culture that flows with the expectations of reason no longer inspired by faith, grace, and authentic conversion.

Finally, postmodernity—regardless of its ambiguous trends and ideas—offers much criticism and hostility toward Christ. We see more and more clearly that the rationalistic type of Christianity that took over Western culture in modern times is fading, revealing a post-Christian or neo-pagan culture that is driven by emotions and ideas very similar to those that dominated pre-Christian era. The tension between Christ and culture is on the increase. In effect, being countercultural becomes not an option but a necessity. Curve **cc** in Diagram 25 ascends with time.

This historical account of Christianity vs. culture provides a broader context for our considerations of the various theories of origins. The fact that current culture strongly favors evolutionism does not mean that Christianity needs to adopt the evolutionary perspective in order to maintain connection with culture. The Christian worldview rests on a synthesis between the permanent and the changeable. If in our times the changeable is greatly favored over the permanent, Christians are entitled to say "no" and go against culture in defending the permanent. This attitude does not mean that Christianity is countercultural or that a believer cannot be a highly cultured person. It only means that sometimes some cultural phenomena need to be challenged by Christians because they stray from the objective and permanent requirements of the faith. The current situation of faith being in opposition to culture resembles that of the early Christianity. And the response of believers today must be similar: people of faith must stand up for the revealed truth because ultimately it is not culture but Christ who saves.

## Chapter III—Summary

In Chapter III we have shown that Christianity used to have a coherent and quite detailed "theory of origins" that was a derivative of the biblical revelation. The Christian concept was challenged by the modern theory of evolution. After Darwin, the original Christian teaching on creation was replaced with theistic evolution—a "theory of origins" that is a blend of Christian teleology and Darwinian evolutionism. Theistic evolutionists believe that their position remains in harmony with modern science while maintaining the basic Christian concept of divine providence acting in nature. However, as we have shown, theistic evolution strays from some of the fundamental tenets of Christianity, such as that in creation no secondary causes take part or that the creative action of God ended for good with the creation of man.

In the first chapter we showed that questions of origins cannot be fully addressed by science. The ultimate answer must come from theology. On the other hand, a simple return to the "young earth creationism" of early Christianity would fail to account for the progress of scientific knowledge that took place in the modern era. For these reasons, we propose a new science-faith synthesis that consists of intelligent design on the level of science and progressive creation on the level of theology. The two levels—science and religion—are harmonized by classical philosophy.

# Summary and Conclusion

THROUGHOUT THIS BOOK WE have presented the basic ideas regarding the question of origins. We have elaborated upon a conflict between two paradigms in today's culture—one naturalistic and evolutionary, the other biblical, based on the belief in creation. Many contemporary Catholics think that Christianity should accommodate itself to the evolutionary paradigm, because it represents the flow and progress of culture. In our opinion, the evolutionary paradigm is precisely what Christianity cannot adopt. The challenge to biological macroevolution presented in this book may be falsely interpreted as a challenge to modern science or culture. But our proposal of the new approach to the dialogue of science and faith is actually a work of a synthesis that is far from challenging culture as such. It is aimed at the renewal of culture, by restoring the central place of the truth about origins both in science and in theology.

The three chapters of this book have guided us through three aspects of the creation–evolution debate. We began our journey by presenting the structure of human knowledge according to the three levels—theological, philosophical, and scientific. Then, we focused entirely on relations between faith and science, according to a systematic as well as a historical approach. We showed that the progress of science has its limits, and that the scientific method cannot address questions of origins. Many issues concerning the operation and structure of the universe are explained by science, but there are some areas which—by the very nature of the inquiry—should be reserved for theology and philosophy. This limitation of science applies primarily to the origin and the destination of the universe.

In the second chapter, our goal was to show what kind of difficulties biological macroevolution encounters from natural knowledge, i.e., science and philosophy. We juxtaposed two theories of origins—(neo-)

Darwinism and intelligent design, presenting the philosophical and scientific elements in both of them. Then we explained intelligent design and how it meets the demands of science. Our conclusion was that intelligent design is a scientific theory that better accounts for the phenomena of nature and thus should replace neo-Darwinism.

In the third chapter, we moved from the natural level of knowledge to the theological level, which involves supernatural premises coming from the Bible and Church teachings. The main goal of that part was to recover the original Christian understanding of creation and to demonstrate how it differs from the contemporary "new creation story" in the form of theistic evolution. Finally, we described three different models of the science–faith synthesis that are present in contemporary Christianity. Our conclusion is that today Christianity is ready for a great renewal of its understanding of origins. All levels of knowledge—traditional theology, sound philosophy, and the best of science—come together in a new synthesis. At the level of theology, this new synthesis adopts progressive creation as the best interpretation of Genesis in accordance with modern scientific data. At the level of science, it adopts intelligent design as the best interpretation of new discoveries in biology. We have shown that there is a perfect compatibility between progressive creation and intelligent design, and that additionally this new synthesis remains in harmony with classical Christian theology and philosophy.

Indeed, only now, having learnt from science about the history of the universe, can we propose a new paradigm for culture and the new faith and science synthesis. The goal of this new synthesis is not to diminish modern science, but rather to reconnect science with reality by stripping it of its layers of reductive philosophy such as Darwinism. This opens the way to reconnecting Christian faith with science without compromising the veracity of the Bible and the traditional teaching on creation.

# Bibliography

Ajdukiewicz, Kazimierz. "Paradoksy Starozytnych," *Filomata* 35 (1931) 6–14, 36 (1931) 51–58.
Aquinas, Thomas. *Scriptum super Sententiis*. Liber 2. http://www.corpusthomisticum.org/iopera.html.
———. *Summa contra Gentiles*. Translated by Anton C. Pegis et al. https://isidore.co/aquinas/ContraGentiles.htm.
———. *Summa Theologiae*. Translated by Fathers of the English Dominican Province. https://isidore.co/aquinas/summa/index.html.
Aristotle. *Metaphysics*. Translated by W. D. Ross. http://classics.mit.edu/Aristotle/metaphysics.html.
———. *Physics*. Translated by R. P. Hardie and R. K. Gaye. http://classics.mit.edu/Aristotle/physics.html.
Artigas, Mariano, et al.*Negotiating Darwin: The Vatican Confronts Evolution, 1877–1902*. Baltimore: Johns Hopkins University Press, 2006.
Asher, Robert J. *Evolution and Belief: Confessions of a Religious Paleontologist*. Cambridge: Cambridge University Press, 2012.
Augustine. *Confessions*. Translated by Henry Chadwick. Oxford: Oxford University Press, 2008.
———. *The Literal Meaning of Genesis*. In *Saint Augustine on Genesis*, translated by E. Hill, 155–506.New York: New City Press, 2002.
Austriaco, Nicanor, et al. *Thomistic Evolution: A Catholic Approach to Understanding Evolution in the Light of Faith*. Tacoma, WA: Cluny Media, 2016.
Axe, Douglas. "Darwin's Little Engine That Couldn't." In *Science and Human Origins*, edited by Ann Gauger et al., 31–43. Seattle: Discovery Institute Press, 2012.
———. "The Case Against a Darwinian Origin of Protein Folds." *BIO-Complexity* 2010, no. 1 (2010) 1–12. https://bio-complexity.org/ojs/index.php/main/article/view/BIO-C.2010.1/BIO-C.2010.1.
———. "Estimating the Prevalence of Protein Sequences Adopting Functional Enzyme Folds." *Journal of Molecular Biology* 341 (2004) 1295–1315.
———. "Extreme Functional Sensitivity to Conservative Amino Acid Changes on Enzyme Exteriors." *Journal of Molecular Biology* 301, no. 3 (September 2000) 585–95.
———. "The Limits of Complex Adaptation: An Analysis Based on a Simple Model of Structured Bacterial Populations." *BIO-Complexity* 2010, no. 4 (2010) 1–10. https://bio-complexity.org/ojs/index.php/main/article/view/BIO-C.2010.4/BIO-C.2010.4.

———, and Ann Gauger. "The Evolutionary Accessibility of New Enzyme Functions: A Case Study from the Biotin Pathway." *BIO-Complexity* 2011, no. 1(2011) 1–17. https://bio-complexity.org/ojs/index.php/main/article/view/BIO-C.2011.1/BIO-C.2011.1

Barbour, Ian. *When Science Meets Religion*. San Francisco: Harper, 2000.

Barr, Stephen M. "The Design of Evolution." *First Things* 156 (October 2005) 9–12.

Beckwith, Francis J. "Intelligent Design, Thomas Aquinas, and the Ubiquity of Final Causes." The BioLogos Foundation, 2010. Available at https://www.scribd.com/document/124769393/Intelligent-Design.

Behe, Michael. *Darwin's Black Box: The Biochemical Challenge to Evolution*. New York: The Free Press, 1996.

———. *Darwin Devolves: The New Science About DNA That Challenges Evolution*. New York: HarperOne, 2019.

———. "Irreducible Complexity: Obstacle to Darwinian Evolution." In *Debating Design: From Darwin to DNA*, edited by William Dembski and Michael Ruse, 352–70. Cambridge: Cambridge University Press, 2004. The chapter is available by itself, with page renumbering, at https://www.lehigh.edu/~inbios/Faculty/Behe/PDF/Behe_chapter.pdf.

———, and David W. Snoke. "Simulating Evolution by Gene Duplication of Protein Features That Require Multiple Amino Acid Residues." *Protein Science* 13 (2004) 2651–64. https://onlinelibrary.wiley.com/doi/epdf/10.1110/ps.04802904.

Benedict XVI. Address to Members of The International Pontifical Theological Commission. December 2, 2011. https://w2.vatican.va/content/benedict-xvi/en/speeches/2011/december/documents/hf_ben-xvi_spe_20111202_comm-teologica.html.

Bolin, Michael J. *"And Man Became a Living Being": The Genesis of Substantial Form*. A lecture delivered at Wyoming Catholic College, October 25, 2013. https://sancrucensis.files.wordpress.com/2015/01/and-man-became-a-living-being.pdf.

Cajori, Florian, ed. *Sir Isaac Newton's Mathematical Principles of Natural Philosophy*. Translated into English by Andrew Motte in 1729. The translations revised, and supplied with an historical and explanatory appendix. University of California Press, 1934.

*Catechismus ex Decreto Concilii Tridentini*. Editio Stereotypa Quinta. Lipsiae: Ex Officina Bernhardi Tauchnitz, 1856. English translation available at http://www.catholicapologetics.info/thechurch/catechism/ApostlesCreed01.shtml.

Chaberek, Michael. *Aquinas and Evolution*. 2nd edition. British Columbia: The Chartwell Press, 2017.

———. Aquinas and Evolution (website). https://aquinasandevolution.org/.

———. *Catholicism and Evolution: From Darwin to Pope Francis*. Kettering, OH: Angelico Press, 2015.

———. *Thomas Aquinas and Theistic Evolution*. http://epsociety.org/userfiles/art-Chaberek%20%28AquinasEvol-Final%29.pdf.

———. "Thomas Aquinas on Creation, and the Argument for Theistic Evolution from *Commentary on Sentences*, Book II." http://epsociety.org/userfiles/Chaberek%20on%20Aquinas%20and%20Creation.pdf.

Darwin, Charles. *The Descent of Man*, London: John Murray, 1871.

———. *On the Origin of Species*. London: John Murray, 1859.

Darwin, Erasmus. *Zoonomia or The Laws of Organic Life*. 2nd edition. Vol. 1. London: J. Johnson, 1794.

De Koninck, Charles. "The Cosmos. The Philosophic Point of View." In *The Writings of Charles De Koninck*, edited and translated by R. McInerny, 256–321. Vol. 1. Indiana: University of Notre Dame Press, 2008.

Dembski, William. *The Design Revolution*. Downers Grove, IL: InterVarsity Press, 2004.

———. "The Explanatory Filter: A Three-part Filter for Understanding How to Separate and Identify Cause from Intelligent Design." An excerpt from a paper presented at the 1996 Mere Creation conference, originally entitled "Redesigning Science." http://www.arn.org/docs/dembski/wdexplfilter.htm.

Denham, Woodrow W. "Two Millennia of Natura Non Facit Saltum Is Enough." https://www.semanticscholar.org/paper/Two-Millennia-of-Natura-non-Facit-Saltum-is-Enough-Denham/4fa7d38361127d8befeaf6ee4ead75769d5f9c7c.

Dennett, Daniel. *Darwin's Dangerous Idea*. New York: Simon and Schuster, 1995.

Dorlodot, Henri. *Darwinism and Catholic Thought*. Translated by Ernest Messenger. London: Burns, Oates, and Washbourne, 1922.

Encyclopedia Americana. "Consonants." *Encyclopedia Americana*. Vol. 3. Philadelphia: Blanchard and Lea, 1857.

Evolution News. "Why the Royal Society Meeting Mattered, in a Nutshell." In *Evolution News & Science Today*, December 5, 2016. Discovery Institute Center for Science and Culture. https://evolutionnews.org/2016/12/why_the_royal_s/.

Feser, Edward. "The Trouble with William Paley." Edward Feser (website), November 4, 2009. http://edwardfeser.blogspot.com/2009/11/trouble-with-william-paley.html.

Fifth Lateran Council. Session 8. 19 December 1513. http://www.intratext.com/IXT/ENG0067/_P9.HTM. Full Council document at:http://www.papalencyclicals.net/Councils/ecum18.htm.

First Vatican Council. Dogmatic Constitution on the Catholic Faith. Chapter 2, "On Revelation." English text from https://www.ewtn.com/catholicism/library/first-vatican-council-1505.

Francis. *Evangelium Gaudii*.24 November 2013. http://www.vatican.va/content/francesco/en/apost_exhortations/documents/papa-francesco_esortazione-ap_20131124_evangelii-gaudium.html.

Galileo. "Letter to the Grand Duchess Christina." In *The Essential Galileo*, ed. and trans. Maurice A. Finocchiaro, 109–45. Indianapolis, IN: Hackett, 2008.

Gilson, Etienne. *From Aristotle to Darwin and Back Again*. Translated by J. Lyon. Notre Dame: Notre Dame University Press, 1984.

Gould, Stephen Jay. *Rocks of Ages: Science and Religion in the Fullness of Life*. Ballantine Books, 1999.

Graney, Christopher M. *The Inquisition's Semicolon: Punctuation, Translation, and Science in the 1616 Condemnation of the Copernican System*. Available at: http://arxiv.org/abs/1402.6168.

Heller, Michael. *Filozofia przypadku* [*Philosophy of Chance*]. Krakow: Copernicus Center Press, 2012.

———, and Jozef Zycinski. *Dylematy ewolucji* [The Dilemmas of Evolution]. Tarnow: Biblos, 1996.

Horn, Stephan Otto, and Siegfried Wiedenhofer, eds. *Creation and Evolution: A Conference with Pope Benedict XVI in Castel Gandolfo*. San Francisco: Ignatius Press, 2008.

Hoyle, Fred, and Chandra Wickramasinghe. *Evolution from Space: A Theory of Cosmic Creationism*. New York: Simon and Schuster, 1984.

Hutton, James. *Theory of the Earth*. Vol. 1. Edinburgh, 1795.
International Theological Commission. *Communion and Stewardship: Human Persons Created in the Image of God*. 2004. http://www.vatican.va/roman_curia/congregations/cfaith/cti_documents/rc_con_cfaith_doc_20040723_communion-stewardship_en.html.
Iverach, James. *Christianity and Evolution*. New York: Thomas Whittaker, 1894.
John Paul II. Address to the Pontifical Academy of Sciences on the subject "Faith Can Never Conflict with Reason." In *L'Osservatore Romano* 44 (1264), 4 November 1992. http://www.elabs.com/van/JPII-faith_can_never_conflict_with_reason.htm.
———. Address to the Pontifical Academy of Sciences. 22 October 1996. http://www.academyofsciences.va/content/accademia/en/magisterium/johnpaulii/22october1996.html.
———. *Fides et Ratio*. Encyclical. 14 September 1998. http://w2.vatican.va/content/john-paul-ii/en/encyclicals/documents/hf_jp-ii_enc_14091998_fides-et-ratio.html.
———. "In Creation God Calls the World into Existence from Nothingness." Catachesis. 29 January 1986. http://www.totus2us.co.uk/teaching/jpii-catechesis-on-god-the-father/in-creation-god-calls-the-world-into-existence-from-nothingness.
Kemp, K. W. "God, Evolution, and the Body of Adam." *Scientia et Fides* 8, no. 2 (2020) 139–72.
Kinkead, Thomas L. *Baltimore Catechism No. 4: An Explanation of The Baltimore Catechism of Christian Doctrine*. http://www.gutenberg.org/cache/epub/14554/pg14554-images.html.
Krapiec, M. A. *Wprowadzenie do filozofii* [*Introduction to Philosophy*]. Lublin: RW KUL, 1996.
Kuhn, Thomas S. *The Structure of Scientific Revolutions*. Chicago: University of Chicago Press, 1962.
Lenartowicz, P. "O 'cudach' probabilistycznych, czyli fakt selekcji i odmowa poznania tego faktu" ["On probabilistic 'miracles,' that is, the fact of selection and refusal of recognition of this fact"]. In *Vivere et Intelligere: Wybrane prace Piotra Leartowicza SI z okazji 75-lecia jego urodzin* [*Selected Works of Piotr Lenartowicz, SJ, on the Anniversary of His 75th Birthday*], edited by J. Koszteyn, 569–608. Kraków: Wyd. WAM, 2009.
Leo XIII. *Providentissimus Deus*. 18 November 1893. http://w2.vatican.va/content/leo-xiii/en/encyclicals/documents/hf_l-xiii_enc_18111893_providentissimus-deus.html (9 November 2015).
Linnaeus, Carl. "De Peloria." Extract from *Amoenitates Academicae* . . . [Leiden: Cornelius Haak, 1749]. In *Issues in Creation, Number 4: Christian Perspectives on the Origin of Species*, ed. P. A. Garner, 7–22. Eugene, OR: Wipf and Stock, 2009.
———. *Fundamenta Botanica*. Amsterdam, 1736.
———. *Philosophia Botanica*. Stockholm, 1751.
Maddox, John. "Down with the Big Bang." *Nature* 340 (10 August 1989) 425.
Maritain, Jacques. "Toward a Thomist Idea of Evolution." In *Untrammeled Approaches*. Vol. 20 of *The Collected Works of Jacques Maritain*. South Bend: University of Notre Dame Press, 1977.
Mayr, Ernst. *Systematics and the Origin of Species*. New York: Columbia University Press, 1942.
McKeough, Michael John. "The Meaning of the *Rationes Seminales* in St. Augustine." PhD dissertation. Washington, DC: Catholic University of America, 1926.

McNew, Sabrina M., et al. "Epigenetic Variation between Urban and Rural Populations of Darwin's Finches," *BMC Evolutionary Biology* 17 (2017), article number 183. https://bmcecolevol.biomedcentral.com/articles/10.1186/s12862-17-1025-29.
Messenger, Ernest C. *Evolution and Theology: The Problem of Man's Origin*. New York: Macmillan Company, 1932.
Meyer, Stephen C. *Darwin's Doubt: The Explosive Origin of Animal Life and the Case for Intelligent Design*. HarperOne, 2013.
———. *Signature in the Cell: DNA and the Evidence for Intelligent Design*. New York: HarperOne, 2009.
Mivart, St. George Jackson. *On the Genesis of Species*. New York: D. Appleton and Company, 1871.
Moran, Laurence. "Evolution Is a Fact and a Theory." The Talk Origins Archive. Updated January 22, 1993. http://www.talkorigins.org/faqs/evolution-fact.html.
Niebuhr, H. Richard. *Christ and Culture*. San Francisco: Harper, 2001.
Oxford Lexicon. "Paradox." *Oxford Lexicon* online. https://www.lexico.com/en/definition/paradox.
Paley, William. *Natural Theology*. Boston, 1854.
Paul VI. Remarks at a symposium on original sin held in Rome on July 11, 1966. *Acta Apostolicae Sedis* 58 (1966), 649–55.
Peacocke, Arthur. "Welcoming the 'Disguised Friend'—Darwinism and Divinity." In *Intelligent Design Creationism and Its Critics: Philosophical, Theological, and Scientific Perspectives*, edited by Robert T. Pennock, 471–86. Cambridge, MA: MIT Press, 2001.
Pelagius I. *Humani generis*. Letter. In *Enchiridion Symbolorum Definitionum et Declarationum*. 34th edition. Herder, 1967.
Pius XII. Address to the Pontifical Academy of Sciences. November 30, 1941. *Acta Apostolicae Sedis* 33 (1941). English translation from *Papal Addresses to the Pontifical Academy of Sciences 1917–2002*. Vatican City: The Pontifical Academy of Sciences, 2003.
———. *Humani Generis*. 12 August 1950. http://w2.vatican.va/content/pius-xii/en/encyclicals/documents/hf_p-xii_enc_12081950_humani-generis.html.
Pohle, Joseph. *God the Author of Nature and the Supernatural (De Deo Creante et Elevante)*. Translated and edited by Arthur Preuss. 2nd edition. St. Louis: B. Herder, 1916.
Polkinghorne, John. *The Faith of a Physicist: Reflections of a Bottom-Up Thinker*. Princeton: Princeton University Press, 1994.
———. *Science and Creation*. Templeton Foundation Press, 2006.
Provinciae Coloniensis. *Acta and Decreta Concilii Provinciae Coloniensis*. Coloniae, 1862.
Russell, Robert John. "Miracles and Science: A Third Way." BioLogos (website). May 16, 2016. https://biologos.org/articles/series/divine-action-a-biologos-conversation/miracles-and-science-a-third-way.
Ryland, Mark. "What Is Intelligent Design Theory?" *Second Spring* 15 (2011).
Santillana, Giorgio de. *The Crime of Galileo*. Chicago: University of Chicago Press, 1955.
Schönborn, Christoph. "Finding Design in Nature." *The New York Times*, July 7 2005, section A, 23. Online version at http://www.nytimes.com/2005/07/07/opinion/07schonborn.html?_r=0.

Siegel, Ethan R. "Ask Ethan: How Large Is the Entire, Unobservable Universe?" Ask Ethan column. *Forbes* online. https://www.forbes.com/sites/startswithabang/2018/07/14/ask-ethan-how-large-is-the-entire-unobservable-universe/?sh=3fad96edf806.

Teilhard de Chardin, Pierre. *Man's Place in Nature: The Human Zoological Group*. London: Fontana Books, 1971.

———. *The Phenomenon of Man*. Translated by B. Wall. New York: Harper, 1959.

Van Till, Howard. "Basil, Augustine, and the Doctrine of Creation's Functional Integrity," *Science & Christian Belief* 8, no. 1 (1996) 21–38.

Weikart, Richard. *From Darwin to Hitler: Evolutionary Ethics, Eugenics, and Racism in Germany*. New York: Palgrave Macmillan, 2005.

———. *Hitler's Religion: The Twisted Beliefs that Drove the Third Reich*. Washington, DC: Regnery History, 2016.

Wells, Jonathan. *The Myth of Junk DNA*. Seattle: Discovery Institute Press, 2011.

Whitcomb, John C., and Henry M. Morris. *The Genesis Flood: The Biblical Record and its Scientific Implications*. Philadelphia: Presbyterian & Reformed, 1960.

Whitehead, Alfred North. *Process and Reality*. New York: The Free Press, 1978.

Zahm, John A. *Evolution and Dogma*. Chicago: D. H. McBride & Co., 1896.

# Index

Abelard, Peter, 264
absolute, 7, 27, 37, 54, 151, 169, 172, 182, 212, 241, 264
   the Absolute, 182, 237
accidental change, 116, 118–19, 121–23, 126, 180, 198, 220
accidental form, 117, 118, 213
Agassiz, Louis, 197
agnosticism, 18, 20, 187
Ajdukiewicz, Kazimierz, 96–97
Ambrose, Saint, 204, 219
anthropology, 33
archeology, 16, 231
Aristotle, Aristotelian, ix, 24–26, 28, 32, 43, 59, 82, 115, 124, 129, 176, 179, 182–83, 226, 233, 235, 237–41, 263
atheist, atheistic, x, 2, 56, 72, 82, 169, 171, 173, 196, 200–202, 232, 235, 242, 270
atheistic evolution, *see* evolution
Augustine, Saint, 68, 90–91, 204, 217–21, 257, 266
Averroes, 26
Avicenna, 26, 222

Baronius, Caesar, 67
beginning, 19, 23, 30, 46, 49, 52, 55, 75–78, 81–83, 88, 92, 99, 108, 125, 136, 138, 162, 167, 169, 178, 182–84, 188–91, 195–96, 203–6, 216, 218, 221, 225, 238, 250–252, 261–62, 272

Behe, Michael, 86, 94–95, 102, 110, 112, 130, 141, 143, 147, 161–62, 224, 252, 254
being, 3, 6–7, 9, 18, 73, 78, 101, 110, 116–17, 119, 121, 123, 124–28, 168, 176, 182–83, 205–6, 209–10, 212–13, 215, 220–221, 227, 236–37, 241, 250, 257–58
   composite, 117, 213
   created, 116, 125, 206, 209–10
   first, 237
   human, ix, 2, 67, 125, 172–73, 182, 246, 255, 270
   infinite, 209
   intelligent, 178–79, 227
   living, nonliving, 56, 66, 68, 74, 80, 88–89, 94–95, 100–101, 109, 121, 125–26, 128, 144, 146, 180, 189–90, 195, 197, 198, 210, 240–242
   material, 3, 7, 106, 116–17, 123, 127, 168, 176, 206, 213, 215
   perfect, 218
   physical, 176
   real, 114, 130
   supernatural, 200
   Supreme, 182–83, 255
Benedict, xiv, Pope, 60
Benedict, xvi, Pope, 159, 248, 251
Bible, ix, 11, 19–20, 61–78, 80, 82–83, 171, 174, 182, 185, 193, 201, 204–5, 211, 219, 225–26, 249–51, 253, 259, 262–63, 276
Big Bang, 12, 39, 52–53, 81, 169, 178, 182, 190, 196, 251–52, 260
Boyer, Charles, 203

Buffon, *see* Leclerc, Georges Louis,
Buridan, John, 27
Butler, Joseph, bishop, 265

Carroll, William, 217
causality, 127, 171, 184, 199, 216
   direct, 171, 174, 216
   divine, 78, 170–71, 185, 191, 200, 207, 216, 220, 224–25, 230, 248, 256
   final, 176
   instrumental 191
   intelligent, 225
   natural, 205
   supernatural, 171, 197, 200, 205, 216
causation, 175, 179, 195, 225
   direct, indirect, 182, 187, 190, 205, 214
   divine, 74, 79, 171–72, 183–84, 190, 192, 200, 216, 220
   intelligent, 177–79, 227, 232, 249, 251, 255
   natural, 74, 171, 200, 230
   material, 175
   primary, 191
   secondary, 183–85, 189–91, 200, 205, 225
   supernatural, 31, 76–77, 170–71, 174–75, 179, 187, 199–200, 220, 225
cause, 7, 30, 55, 72, 95, 115, 125–29, 144, 146, 151, 153, 158, 162, 168, 176–79, 182, 184, 187, 195, 214–15, 231–32, 235, 256
   active, 222
   efficient, 128, 176–78
   final, 56–57, 127–29, 131, 176–77, 188, 200
   first, 6, 24, 31–32, 168, 183, 209, 241, 250
   formal, 128–29, 176
   immaterial, 177
   instrumental, 128, 209
   intelligent, 169, 174, 177, 178–80, 232
   lower, 31, 125
   material, 127–29, 174–78,
   natural, 74, 77, 171–72, 175, 180, 193, 215, 220, 229
   passive, 211
   physical, 230
   prime, 128
   proper, 6
   secondary, 74, 76, 88, 127, 183–84, 187, 189, 191, 196, 209–10, 212, 214, 217, 220, 222, 230, 250, 274
   sufficient, 123, 126, 198
   supernatural, 175, 177, 180, 183, 229–31
   true, 238
   ultimate, 6–7, 11, 32, 83, 168, 236
Caverni, Raffaello, 62, 196, 245
certitude, 8–9, 66
chance, ix, 31, 57, 120, 126–27, 146–53, 156–59, 194, 201, 221–23, 227, 231–32, 249–52, 254
chaos, 53, 55, 124, 194, 201, 206, 213
Christ, Jesus, 5, 19–20, 182, 228, 246, 259–60, 262–73
Christina of Tuscany, Duchess, 63, 70
Clement of Alexandria, Saint, 25, 265
Clement of Rome, Saint, 25
Columbus, Christopher, 270
common ancestry, 56, 98–99, 139–41, 146, 163, 190, 220, 225, 252, 256, 259
complexity, 86, 102, 110, 124, 160–62, 223–24
   irreducible complexity, 160–62
   specified complexity, 151
Comte, August, 20–22, 31, 233
conflict, 1–3, 12, 14, 22, 24, 70, 72, 74, 81, 127, 196, 233, 236, 238–39, 241, 258–59, 265, 275
Copernicus, Nicolaus, 28, 39, 59, 73, 79
Coyne, George SJ, 251
Coyne, Jerry, 187
creation, 5, 11–12, 14, 17, 19–20, 34, 56, 58, 62–63, 67–68, 75, 77–79, 81–82, 88, 90, 122, 126–28, 136, 147, 161, 168, 170, 178–79, 181–84, 188–95, 197–200, 202–4, 206–21, 223–25, 250–253, 255–64, 266, 268, 274–76
   continual, 213
   history of, 19–20, 69, 191, 199, 204, 207, 223, 225, 259

# Index

progressive, 186, 190–91, 193, 201–2, 204, 252, 256, 274, 276
special, 79, 184, 191–92, 195, 197–200, 241, 245–48
theology of, 16, 256–60
creationism, 167–68, 171, 191–94, 201–2, 260
young earth, 79, 186, 191, 193, 200–202, 274
Creator, 11, 30, 160, 168, 178, 184–85, 195–96, 236, 260, 264–65
Crick, Francis, 37
cultural, *see* culture
culture, x, 23–24, 26–29, 31–32, 34, 62, 66, 71, 73, 91, 114, 123, 181–82, 184–85, 201, 208, 230, 233–36, 257–58, 260–261, 263–73, 275–76

Darwin, Charles, ix, 3, 31, 58, 62, 73–74, 80, 82–83, 85–87, 90–93, 95–111, 113–15, 118, 123, 126, 130, 133–36, 138–39, 141–42, 160–61, 165, 170–71, 184–85, 187, 190, 195–96, 203, 217, 221, 224, 243–46, 255, 258, 262, 273
Darwin, Erasmus, 30, 184
Darwinism, 31, 34, 96, 146, 167, 169–72, 174–75, 229, 244–45, 261, 276
data, x, 1, 10, 16, 37, 39–42, 44, 47, 53, 60, 67–68, 104, 132, 134, 146, 182, 193, 200, 208, 219, 249, 252, 254–55, 259–60, 262, 276
De Koninck, Charles, 197, 241
deism, 30, 184, 193–95
Descartes, René, 27–28
design, *see* intelligent design
destination, xi, 6, 71, 84, 182, 258, 275
dialogue, 1, 13–15, 32, 59, 61–63, 95, 115, 250, 275
discovery, 30, 36–37, 39, 49, 70, 79, 103–4, 183, 228, 239, 270
divine action, 69, 94, 193–94, 207, 214, 255
dogma, *see* dogmas
dogmas, 9, 16, 59, 184, 251, 258, 263
Dorlodot, Henri, 196, 243

Einstein, Albert, 42, 50, 147
Emerson, Ralph Waldo, 264
empirical, 2, 6, 8, 10, 18, 41–42, 48, 52, 87, 101, 127, 167, 184, 190, 195, 235
Enlightenment, 28–30, 184
entropy, 53, 55, 124, 213
epigenetics, 106, 143–45, 147, 254
essence, 60, 65, 116–17, 210, 225, 239
Eubulides, 96–98, 113, 123
evolution, ix–xi, 4, 14, 19, 22, 56–57, 62–63, 67, 82, 85–91, 93–96, 101, 107, 110, 115, 117, 123, 125–32, 138, 142–44, 146–48, 159–61, 163–64, 181, 185, 189–90, 193–97, 200, 202–3, 209, 214, 217, 221, 224–25, 241–48, 250–251, 255, 257, 273, 275
biological, 56, 91–92, 187, 218, 223
Darwinian, 113, 127–28, 147–48, 164, 223, 244, 250
macroevolution, 1, 16, 54, 56, 63, 66, 68–71, 85–86, 92–96, 105, 113–15, 118–19, 121–23, 125–26, 128, 130–32, 143, 145, 171, 180–81, 187, 190, 202–3, 239–43, 251–52, 254, 262, 275
materialistic, 125, 186–87, 200, 249–50, 256
microevolution, 68, 92–94, 112, 118, 224
polyphyletic, 197
theistic, 56–58, 118–20, 127–29, 171, 179, 181, 184–91, 193–97, 200–201, 204, 208, 216–17, 219–21, 223–26, 243, 248, 250, 254–56, 258–60, 273–74, 276
atheistic, 56, 129, 171, 186, 200, 245
evolutionists, 57–58, 62, 68–69, 82, 112, 119, 123, 129–31, 171, 179, 181, 185, 187, 189–90, 194–96, 200–202, 209, 213–14, 216–17, 219–25, 241, 243–44, 249, 259–60
experiment, 2, 6, 26, 29, 36, 47, 50, 56, 60, 106, 121–22, 144, 163, 167–68, 237–38, 254
explanatory filter, 149–51, 153–58, 178, 232

fact, ix–x, 5–6, 10, 12, 14–15, 19, 21, 30, 39, 41–42, 46–47, 57, 59–62, 66–68, 70–71, 74, 79–81, 85, 88, 91, 94, 102, 104, 107, 110, 113–14, 119, 123–25, 132, 134–35, 138, 140–44, 147, 150, 156–58, 162, 164–65, 168, 170–72, 177–78, 180, 183, 190, 192, 194–96, 200–202, 212–13, 215, 218, 221, 223–24, 228, 231, 233, 238, 243–46, 248–55, 257, 259, 267–68, 273

faith, x, 2, 11, 13–17, 22, 25–26, 28, 57, 61–64, 79, 127, 178, 180–84, 192–93, 196, 201, 205, 228, 236, 244, 246, 249–50, 253–60, 262–63, 265, 267, 269, 271–76

fideism, 2, 13, 16, 69, 200, 201, 259,
fideistic, *see* fideism
finality, 127, 176
finite, 36, 38, 40–43, 45–48, 52, 82
  universe, *see* universe
finitude, 43–46
fixism, *see* species fixism
foresight, 131–32, 146, 162, 178
form, 2, 4, 6, 9, 17–19, 22, 34, 45, 53–56, 59, 74, 76–77, 79, 82, 86, 88–93, 95–96, 103–8, 110, 112, 116, 119–20, 122, 125–26, 128–29, 131–34, 140, 144–45, 155, 165–67, 174, 176–77, 182, 184–88, 193–96, 199, 201, 204, 206, 210–213, 217–24, 238, 240–242, 244, 249, 251, 256, 261, 266, 276
  accidental, 117–18, 213
  individual, 119, 121
  substantial, 88–89, 118–22, 128, 180, 198, 241
  transitional, 105, 107–8, 133–34
Freud, Sigmund, 31
fundamentalism, 13

Galileo, Galilei, 3, 28, 58–71, 73, 79, 83, 233, 238
Gauger, Ann, 163–64, 252
genera, *see* genus
Genesis, 62, 66–68, 71, 74–78, 80, 82, 89, 124, 126, 183, 185, 190, 197–98, 203–6, 208, 210–212, 218–19, 221, 225, 245, 247–48, 250, 252–56, 260–263, 276
genus, 86–87, 89, 92, 98–99, 102–3, 136
geocentrism, 62, 65
geology, 16, 30, 50, 74, 76–78, 260
God, ix–x, 1, 2, 5, 7–9, 11–20, 24–25, 30, 34, 47–48, 55–58, 63, 66–68, 70, 72–79, 82, 91, 116, 118–20, 125, 127–29, 171–74, 176, 182–95, 197–201, 203–15, 217–18, 220–232, 234, 236–37, 241–42, 246–47, 249–56, 258–62, 263, 265–66, 269, 271–72, 274
  of the gaps, 147, 194, 229–32
Gregory XIII, Pope, 59
Grotius, Hugo, 28

happiness, 7, 22–23, 34, 270
happy, 8, 36
Hartmann, Nikolai, 265
Hegel, Georg Wilhelm Friedrich, 21, 264
Heisenberg's uncertainty principle, 152
heliocentrism, 28, 58–59, 63, 67, 270
Heller, Michael, 185, 188, 217, 251
hermeneutics, 63
  hermeneutical principle, 65–67, 69–70, 72–74
Herschel, William, 39
history, 3, 12, 16–17, 19–22, 27, 30, 42, 50, 55–58, 69, 71, 73, 75, 79, 86, 89, 101, 130, 135–38, 144, 156–57, 164, 169–71, 173–74, 182–85, 189, 191, 196, 198–99, 201, 204, 206–7, 211, 214, 219, 223–26, 244, 251–52, 255, 259, 265, 272, 276
Hobbes, Thomas, 28
Holy Scriptures, 5–6, 64–65, 67, 73, 77
homology, 138–42
Hoyle, Fred, 148
Hubble, Edwin, 39
Hume, David, 28
Hutton, James, 30
hypothesis, 30, 34, 44, 59, 65, 91, 185, 196, 248, 252

## Index

immaterial, 3, 7, 11–13, 18, 48, 116, 128, 168, 173, 175, 177–78, 211, 240–242
infinite, 30, 36, 38, 41–48, 53, 55, 102, 149, 155, 184, 209
   universe, *see* universe
infinity, 46–47
intelligent design, (ID), 1, 85, 94, 110, 145–47, 149, 152, 163, 167–71, 174–75, 177–81, 202, 226–36, 239–40, 249, 251–52, 255–56, 260, 274, 276
invisible, 5–6, 9–10, 12–13, 17, 64, 67, 112, 195, 197, 201, 236–37, 239, 259
Irenaeus, Saint, 272
irreducible complexity, *see* complexity

Jefferson, Thomas, 264
Jesus Christ, *see* Christ
John Paul II, Pope, 60, 63, 248, 251, 257–58
Justin Martyr, 25

Kant, Immanuel, 10, 28, 264
Kepler, Johannes, 73
Kierkegaard, Soren, 265
knowledge, x, 1–11, 13, 17–18, 21, 23, 25–29, 32–34, 36–37, 45, 48–50, 52–53, 55, 57, 60, 64, 77, 84, 89, 115, 131–32, 139, 156, 167–68, 172, 174, 201, 203, 226, 230, 232–33, 235, 237–39, 249
   absolute, 27
   domain(s), disciplines of, 5, 22, 34, 89, 173, 198, 228–29
   empirical, 2, 184
   form(s) of, 34, 53, 177
   human, x, 3–5, 9, 39, 55, 169, 197, 230, 253–54, 275
   level(s) of, 5, 9–10, 61, 89, 95, 115, 123, 176, 180, 229, 276
   natural, x, 1, 12, 17, 35, 71, 85, 167, 181, 193, 230, 275–76
   positive, 12, 17, 21
   religious, 1, 54
   scientific, 11–13, 35–36, 49, 52–54, 57, 60, 65, 230, 274
   supernatural, theological, revealed 1, 5–6, 11–12, 17, 19–20, 53, 61, 64, 67, 95, 168, 200, 272
   true, 9–10, 14, 18, 22, 28, 56, 238
   type(s) of, x, 2, 4, 11, 17–18, 54–55
   ultimate, 2, 9, 15, 31
Krapiec, Mieczyslaw A., 197
Kuhn, Thomas S., 146–47

Lamarck, Jean-Baptiste, 95
Lamarckism/Lamarckian, 95, 243
Laplace, Pierre-Simon, 30
Laun, Andreas, bishop, 252
laws of nature, (natural laws), 16, 30–31, 55, 69, 79, 83, 160, 171, 184, 186, 214, 221, 223, 230, 232, 255, 258
Leclerc, Georges Louis, 88–89
Leibniz, Gottfried Wilhelm, 28, 101, 264
Lemaître, Georges, 39
Lenartowicz, Piotr, 149–50
Lenin, Vladimir, 154–55
Leo XIII, Pope, 68, 244, 247
Leroy, Dalmase, 62, 196, 245
Linnaeus, Carl, 86, 88–89, 98, 197
Locke, John, 28, 264
logic, 6, 32–33, 85, 96–98, 100–102, 111, 113–14, 122–23, 133, 138, 140, 148, 163, 175, 180, 232, 235
LUCA, (last universal common ancestor), 95
Luther, Martin, 265
Lyell, Charles, 30, 79, 101

macroscale, 36, 39–41, 45, 48–49
magisteria, *see* magisterium,
magisterium, 2, 14, 75, 249, 253
Malebranche, Nicolas, 28
Marcion, 265
Maritain, Jacques, 197
Marx, Karl, 31
material, 3, 5, 7, 11, 18, 30, 39, 43, 45, 62, 82, 106, 116–18, 120, 123, 125–29, 131, 140, 154, 167–68, 170–73, 176–79, 197, 199, 206, 209, 211, 213, 215, 237–38, 240–243, 250, 259
materialism, 18, 179, 187, 232, 240, 245, 259,

materialists/materialistic, 3, 56, 70, 74, 82, 125, 129, 133, 169, 172–73, 177, 179, 186–87, 195, 200–201, 221, 232, 235, 241–43, 249–50, 256, 259
matter, 2, 4, 38–39, 53–55, 64–65, 80–82, 91, 116, 119–22, 124, 126–28, 152, 174–77, 186–87, 195–97, 200–201, 206, 210–212, 214, 220, 222, 231, 236, 240–242, 247–48, 250, 255
   prime matter, 125
Maurice, Frederick D., 266
Mazzella, Camillo, 203
McKeough, Michael, 217
McMullin, Ernan, 217
mechanism, 57, 94, 95, 100, 111, 114, 144–45, 147, 161, 170–72, 174, 180, 241, 243, 251
   biochemical, 240
   biological, 92, 241,
   Darwinian, 111–13, 118, 126, 147, 162, 164, 171, 189–90, 243, 254
   evolutionary/of evolution, 94–95, 118, 132, 171, 184, 190, 202, 221–22, 243, 250, 256
   material, 170, 174
   materialistic, 251
   neo-Darwinian, 94, 106, 144–46, 159–60, 162–63, 167, 198, 254
Messenger, Ernest, 196, 200, 243
metaphysical, *see* metaphysics
metaphysics/metaphysical, 21–22, 24, 27, 33, 67, 85, 87–88, 96, 114–19, 121–30, 179–80, 201, 213, 218, 233, 235, 237–39, 256, 262
method, 2, 4, 6, 13–14, 34–36, 46, 95, 168, 228, 231, 236
   deductive, 29
   demonstrative, 64
   experimental, 29, 44, 175, 233
   philosophical, 6
   rational, 27–28,
   scientific, 6, 17–18, 34–35, 40–42, 44, 49, 54, 58, 152, 176, 178, 180, 193–94, 201, 228–29, 231, 275
   theological, 6
microscale, 36–38, 41, 48–49
Middle Ages, 25–28, 232, 272

mind, 3–6, 9–10, 21–22, 24, 28, 32, 45–48, 59, 63, 70, 73, 76, 81–82, 88, 90–91, 99, 116, 118–19, 121, 124, 128, 130–31, 134–35, 141–42, 152, 155–56, 158, 163, 169, 175, 183, 193, 196, 204, 207–8, 210, 218, 222, 228, 232–37, 240, 242, 247, 249, 252, 260, 262
minimalistic, 21, 33
miracle, 5, 12, 16, 19–20, 30, 55, 58, 72–73, 147, 149–50, 158, 172–73, 183, 189, 193, 202, 207–8, 214, 216, 223, 231
Mivart, St. George, 190, 196, 208–9, 217, 243
model, 1–3, 9–10, 30, 37, 39, 42–43, 47, 59, 65, 79, 81, 109, 124, 131, 152, 169, 175, 181, 218, 224–25, 250–257, 263–67, 272, 276
modern, *see* modernity
   science, *see* science
   geology, 76–77
   biology, 115, 125, 161
modernism/modernist, 22, 263, 265,
modernity/modern, 6, 21–22, 27–29, 49, 62, 86, 88–89, 91, 101, 125, 141, 166, 183–86, 213, 217, 219, 232–33, 236–39, 248, 253, 256, 258, 261–62, 264, 266, 270, 272–74, 276
monism, 17, 187–88, 195
Moses, 204
multiverse, 34, 44, 94, 185, 196, 251–52
mythology, 14, 21, 24–25, 71, 182, 185, 225, 234–36, 260

Napoleon, 30
natural, 2, 6–7, 9, 16, 19, 26–27, 30, 33, 42, 46, 53, 65–68, 72, 74, 77, 79, 86, 125, 128, 132, 148, 168, 171, 173, 185, 189–92, 194, 197, 207, 217, 220, 225, 227, 230, 232–33, 235–38, 247, 269
   cause, *see* cause
   conditions, 48, 86, 161, 199
   explanation, 57–58, 72, 74, 172, 174, 179, 184, 229–32
   events, 65, 72, 94, 128–29, 160, 189, 214, 232

# Index

evolution, *see* evolution
generation, 94, 118, 187, 191, 205, 212, 223, 263
history, 16, 79, 86, 135, 170–71, 198–99, 201, 211, 219, 224, 252, 259
investigations, 18–19
knowledge, *see* knowledge
order, 16, 58, 65, 68, 190, 194, 213–15, 217, 224, 227–28, 230, 250
phenomenon/phenomena, 13, 16, 72, 81, 167–68, 172, 193, 229–30, 233–34, 238, 251
philosophy, *see* philosophy of nature
power, 226
process, 53, 74, 77, 91, 94, 128, 171, 225, 241, 250, 255
reason, 11, 168, 228
resources, 224
selection, 74, 82, 87, 94–95, 99–103, 106, 110–13, 118, 126–28, 138, 140, 144, 146–47, 152, 156, 158–60, 163–64, 167, 170, 180, 190, 221, 243, 254–55
science, *see* science
species, *see* species
theology, 228
thing(s), 207, 227, 233
world/universe, 68, 127, 168, 172, 227, 231–32, 237
natural laws, *see* laws of nature
natural philosophy, *see* philosophy of nature
naturalism, 2, 16–20, 34, 41, 53–54, 56–58, 67, 84, 172–73, 179, 197, 202, 208, 229, 234
methodological (epistemological), 173–75, 177, 179–80, 194, 231
ontological (metaphysical), 179
naturalistic, 69, 170, 196–97, 200–202, 275
nature, ix, 1–3, 6, 11–12, 16–18, 30–31, 41–42, 49, 54–55, 57–58, 60, 62, 64, 66–69, 75, 77–78, 80, 82–84, 86–89, 94, 98–99, 101–4, 110, 112, 114, 116–18, 121, 123–24, 128–29, 131–34, 151–52, 157, 159–60, 162, 168–80, 182, 184, 186–91, 193–95, 198–201, 204–5, 207–15, 218–21, 223–24, 226–28, 230, 232–43, 246, 249, 251–53, 255–57, 261–62, 264–66, 270, 274–76
book of, 11, 174, 249
necessity, ix, 21, 31, 39, 58, 64, 68, 126–27, 146–47, 152, 154–55, 157–59, 221, 227, 231, 249–52, 273
neo-Darwinism, neo-Darwinian, 1, 94–95, 106, 127, 143–47, 152, 156–59, 162–63, 174, 180, 198, 240–241, 250, 254, 256, 262, 276
Newton, Isaac, 30, 50, 60, 73, 101, 147
Niebuhr, Richard, 263–73
Nietzsche, Friedrich, 31
NOMA (non-overlappig magisteria), 2, 14, 194
nominalism, 27, 33, 89, 129
nothingness, 12, 47–48, 52–53

occasionalism, 194, 199–200, 216, 241
origin, *see* origins
origin(s), ix, 1, 13–16, 56, 63, 67–68, 71, 74, 76–81, 83–84, 87, 89–90, 94, 101, 109–11, 130, 136, 146, 160–61, 170–71, 173, 181, 197, 201, 203, 208, 225, 245–46, 250, 253, 256, 275–76
of human/humanity/of man, 16, 66–69, 71, 74–75, 140, 182, 245–48, 253, 258, 261
of species, *see* species
of the universe, 16, 68, 73, 75–76, 81, 83, 173, 182–84, 186, 213, 218, 224, 230, 256, 260
problem of, 15
question of, ix–x, 3, 13–16, 63, 70, 72–73, 84, 133, 173, 179, 206, 223, 253, 258, 274–75
theory of, 3, 16, 66, 169, 181, 184, 273, 275

Palmieri, Dominic, 203
pantheism, 188, 193, 245
paradigm, x, 146–47, 156, 185, 190, 272, 275, 276
biblical, 183
evolutionary, 115, 243–44, 275
naturalistic, *see* naturalistic

# Index

paradox, 38, 43, 96–98, 101, 103, 105, 108–10, 112–14, 122–23, 263, 265
particle, 7, 37–38, 45, 83, 106, 121, 124, 149, 152, 155–56, 159, 175, 240
Paul VI, Pope, 248
Paul, Saint, 206
Perrone, Giovanni, 203
Pesch, Christian, 203
phenomena, 6, 10, 16, 21, 27, 31, 36, 42, 50, 57, 65, 68, 74, 79–80, 96, 115, 124, 134, 147, 152, 159, 172, 174, 230, 266, 270, 276
  cultural, 264, 273
  material, 175
  natural, *see* natural
  physical, 17, 55, 230, 240, 246,
  supernatural, 13
phenomenon, *see* phenomena
philosophy, x, 2–9, 11–12, 20–34, 38, 42–43, 45, 47–48, 50, 53, 55, 61, 84–85, 89, 95–96, 114–15, 124, 129, 132–33, 167–70, 173, 176–80, 185, 188–89, 192, 197, 211, 228–43, 246, 252, 254, 256, 258–60, 262–63, 270, 272, 274–76
  of nature/natural, 26, 129, 168, 176, 232–42, 270
physics, 26, 34, 37, 45, 68, 78, 83, 124–25, 147, 151, 194, 233, 237–39, 260
Pius V, Pope, 75
Pius IX, Pope, 247
Pius XII, Pope, 132, 246–48
Planck time, 51–52
Plato/Platonic, 24–26, 33, 82, 182–83, 226
Pohle, Joseph, 203
positivism, 21, 31–33, 184
postmodern, *see* postmodernity
postmodernity, 3, 22, 32–34, 273
prehistoric, 22–23, 35
process, 34, 40, 53, 56, 77, 79, 91, 99, 101–2, 107, 109–12, 119, 125, 127, 131, 142, 143–44, 152, 159, 162, 171, 184, 188–90, 212, 217, 224, 242, 269
  biological, 241–42
  cosmic, 56

evolutionary/of evolution, 57, 91, 112, 127, 143, 188, 194–95, 200, 209, 223, 241, 246, 250, 254, 256, 261
  geological, 76–78, 101
  natural, *see* natural
  material, 118, 131, 242
  of generation, 152, 171
  of natural selection, 102–3, 110, 112–13
protein, 7, 60, 92, 95, 106, 111–12, 130–31, 144–46, 155–57, 161–64, 194, 224, 242
Ptolemaic (astronomy), 26, 65

rationalism, 17, 28–29, 270
realism, 124, 129, 185, 223, 225, 236, 239
reality, ix–x, 1–3, 6–7, 9–10, 18, 23–24, 28, 34, 41–46, 82, 88, 95, 98–100, 115–16, 124–25, 131, 149, 168, 176, 188, 192–93, 215, 236–39, 249, 263–64, 266, 268–69, 276
  biological, 98, 101–2
  external, 5, 10, 28, 236
  material/immaterial, 18, 82, 125, 237
  physical, 9, 35–36 , 38, 59, 66–67, 83, 89, 140, 169, 188, 199–200, 233
  visible/invisible, 5, 12–14, 19
realm, 2–3, 14, 29, 39, 110, 132–33, 175, 193–94, 234, 265, 266, 268, 270
  biological, 94, 109, 125, 140, 142
  material, 30
  ideal, 116
  natural, 2
  physical, 237
  supernatural, 9
  visible/invisible, 13, 17, 67, 197, 236
reason, 1, 11, 13–15, 24, 26–29, 68, 72, 110, 159, 251, 261, 264, 267, 273
  human, 2, 6–8, 11, 13, 18, 226, 228
  natural, 11, 168, 228
reductionism, 22, 31, 129, 174, 177, 179, 193, 235–36, 239–42
religion, 1–3, 12, 14, 20–29, 31–32, 34, 61–62, 84–85, 91, 172, 177, 182, 201, 236, 264–65, 267–71, 274

revelation, 11, 16, 64, 168, 183, 217, 245,
    249–50, 257, 264–65, 267
  biblical, 273
  Christian, 53, 257
  divine, 7, 13
  private, 168
  special, 207
  supernatural, x, 2, 6, 19, 54, 69, 228,
    237, 249, 255
Ritschl, Albrecht, 264
Ross, Hugh, 253
Rousseau, Jean-Jacques, 28

salvation, 6–7, 20, 22–23, 63, 182, 213,
    215, 260
  history of, 19–20, 57–58, 73, 183,
    189, 207, 214, 224, 259
Schleiermacher, Friedriech, 264
Schönborn, Christof, 252
science, ix–x, 1–9, 12, 18, 20, 22–23,
    25–29, 31, 34–37, 40–42, 44–45,
    47–55, 57–58, 60–67, 69–72,
    75–77, 79–85, 89, 91, 95, 106,
    132–33, 146–47, 151, 157–60,
    167–69, 171–81, 184–85, 189,
    191–98, 201, 203, 227–36, 238–
    39, 243–44, 249–54, 256–60,
    262, 274–76
  and faith, x, 2, 13–17, 57, 181, 249,
    258–59, 275–76
  and theology, 3, 12, 14, 62, 132–33,
    230, 275
  and religion, 1–3, 22, 32, 61, 84–85,
    236, 265, 274
  biological, 123, 146, 148
  experimental, 26, 38, 123, 270
  historical, 60
  limits of, 3, 34–39, 41, 45, 48–49,
    52, 54, 56–57, 69, 78, 84, 172,
    173–74, 177, 248, 253
  modern, 41, 73, 84, 175–76, 233,
    237–39, 252–53, 274, 276
  natural, 2, 4, 6–8, 11, 14, 26–27,
    34–35, 44, 57, 61, 64, 74, 95,
    157, 168–69, 172, 177, 180,
    235–36, 239–40
scientism, 14, 28, 233–34, 236
self-organization, 94, 123–25, 186, 220
seminal reasons, 204, 218–20

Settele, Giuseppe, 60
singularity, 12, 52–53
skepticism, 8, 32, 87
soul, 4, 11–12, 18, 119, 126, 173,
    196–97, 200, 207, 211, 214, 234,
    240–241, 245–47, 261
space, 35–36, 38–43, 45–47, 52–53, 55,
    145, 148, 176, 179, 210, 218, 250
special creation, *see* creation
species, 16, 40, 58, 69, 74, 79, 82, 85–89,
    92, 98–110, 113–14, 122–23,
    126, 128–30, 133–34, 136, 139,
    142, 145, 164–65, 171, 173–74,
    176, 187, 191, 196–98, 205–7,
    210, 213–14, 218–22, 224–25,
    241, 254, 261
  biological, 86–89, 92, 99, 197–98,
    224
  creation/formation/production/
    emergence/generation of, 19, 30,
    58, 68, 79, 100, 102, 125, 127,
    134, 170–72, 174, 190–94, 198,
    201, 206, 210–212, 216, 220,
    222–23, 241–42, 252, 259
  evolution of, 82, 113
  fixism, 88–89, 98, 217
  logical, 87–88
  metaphysical, 88, 119–20, 207, 222
  natural, 82, 88–90, 94–95, 115, 187,
    190, 198, 221–22, 224
  new, 86, 88, 90–95, 98, 100, 102–4,
    108, 110, 112, 122, 126, 131,
    136, 138, 145, 160, 173–74, 189,
    194, 210, 213–14, 222, 242
  notion/idea/definition/concept of,
    85–89, 95, 98, 102, 107, 109
  origin of, ix–x, 1, 14, 16, 56, 57–58,
    68–69, 73–74, 82–83, 87–90, 92,
    95, 97–101, 103, 110–11, 131,
    133–34, 142, 146, 160, 172–74,
    181, 184, 186, 191–92, 195–97,
    200–202, 213, 217, 219–23, 225,
    242–43, 245–46, 250–251, 254,
    258
  transformation/change/stability
    of/limits of, 56, 88, 92, 93–94,
    99–101, 103–4, 109, 113–14,
    118, 122, 129–31, 134, 146, 180,
    190, 220, 252, 254, 256

Spencer, Herbert, 90–91, 123, 142, 217
Spinoza, Baruch, 28
Stackpole, Robert, 252
subjectivism, 28
substance, 26, 88, 116–23, 177, 179, 197, 209–10, 212–13
substantial change, 90, 116, 118–19, 123
substantial form, 88–89, 118–22, 180, 198, 241
supernatural, 2, 5–6, 12, 20, 31, 55, 70, 74, 76, 78, 132, 172, 179, 183, 185, 193, 197, 200, 208, 212, 214, 216, 228, 230–232, 235, 247, 250, 265, 269–70, 276
   action, activity, 30, 69, 72–73, 76–78, 191, 194, 199
   being, *see* being
   causation, *see* causation
   cause, *see* cause
   event(s), 20, 70, 173, 189
   force, 177
   formation, 19, 81–82, 183, 193, 200–201, 211–12, 215, 223, 256
   knowledge, *see* knowledge
   phenomena, *see* phenomena
   power, 38, 48, 188, 200, 229–30,
   realm, *see* realm
   revelation, *see* revelation
   work of God, 58, 78
synthesis, x, 13, 14, 17, 26–27, 57, 61, 144, 146, 180, 236, 249–50, 253–55, 257, 259–60, 262, 265–66, 268, 270–276

Tanquerey, Adolphe, 203
Tertulian
the origin of species, *see* species, the origin of
theistic evolution/evolutionism, *see* evolution
theology, 2–9, 12–14, 16, 18, 20–22, 25–29, 34–35, 39, 48, 55, 57–58, 61–63, 71–72, 74, 81, 84, 89–90, 95, 115, 124, 127, 132–33, 167–69, 172–73, 175, 179–81, 183–84, 189, 192, 194, 197, 200–201, 203, 216, 228, 230, 235–36, 243–44, 248, 253–54, 256–60, 263, 270, 274–76
   of the body, 34, 258, 263
   theory, 62, 67, 69, 76, 80–81, 85, 87, 91, 94, 99–101, 104, 107, 109, 146–47, 149, 156, 160, 167, 169–72, 180–81, 196, 235, 239, 244, 251, 255, 257, 259, 265
   Darwin's/Darwinian/neo-Darwinian, 3, 31, 58, 80, 82, 85, 87, 93, 98–102, 104, 107, 109–10, 114–15, 123, 130, 133, 135–36, 139, 142, 157, 160–61, 167–69, 185, 221, 244–46, 254
   heliocentric/Copernican, 28, 60
   natural, 174
   of biological macroevolution, 62, 85, 123, 126, 130, 171, 180–81
   of creation, 168
   of evolution, *see* evolution
   of intelligent design, *see* intelligent design
   of origins, 3, 16, 184, 273
   scientific, 172, 178, 196, 201, 236, 239, 276
Thomas Aquinas, Saint, 5, 9, 26, 28, 45, 47, 115, 119, 126–27, 129, 151, 183, 195, 198, 204–5, 211–13, 219, 221–24, 226–27, 241, 257, 263, 265
Thomistic evolutionists, 225
Time, 7, 19, 23, 25–26, 30–31, 39–43, 46, 49–53, 55–56, 59, 61–62, 66–71, 73, 86, 90–91, 93–94, 98, 100, 102–4, 106, 108, 132–33, 136–38, 142, 148, 152, 156, 164, 183–86, 188, 191, 196–97, 199, 205–8, 213, 217—219, 225, 232, 234, 236, 238, 243, 248, 250–252, 254–55, 266–73,
   beginning of, 19, 178, 182, 189–91, 195, 205, 218, 250, 252
   deep, 30, 79, 190–91, 193, 202, 260
   end of, 5
   modern, 88, 184, 264, 266, 273
   Planck, 51–52, 81
   prehistoric, 22, 35
   succession, 204, 218
   timeline, 52, 205, 224
   timescale, 30, 50, 191, 201, 220
Tolstoy, Leo, 264

Trinity, 5, 182, 228
truth(s), 9–14, 19–20, 27–28, 33, 48, 51, 57, 59, 61–62, 77, 79, 150, 173, 182, 203, 205–6, 228, 233, 236, 239, 248, 253, 260, 262, 268–69, 271–72, 275
   biblical, 182
   cannot contradict truth, 12, 122–23, 132, 167, 256
   historical, 71, 247, 260
   religious, 57, 70
   revealed, 12, 19–20, 84, 250–252, 273
   scientific, 62, 202
   self-evident, 8
   spiritual, 5
   supernatural, 132, 228, 247
   theological, 16, 19–20, 57–58, 61, 67, 132, 248
   ultimate, 24

universe, ix, 3, 5, 7, 11, 14–17, 25, 30, 34, 39–45, 47–48, 50, 52–59, 68, 73, 75–80, 82, 91, 101, 124, 149, 151, 156–57, 168–69, 172, 176, 182–92, 200–202, 207–8, 213–14, 222–25, 227–28, 230–231, 236–38, 250, 252, 257–58, 264, 271–72, 275
   age of, 78, 80, 252, 256
   beginning of, 12, 23, 52, 88, 182, 184, 196
   creation of, 19, 82, 182–83, 203, 250
   end of, 23, 42, 176
   emergent, 124
   eternal, 82, 225
   expansion of, 45, 81–82
   finite/infinite, 47–48
   formation of, 19, 30, 58, 81–82, 183, 189, 207, 211, 215–16, 223–25, 250, 252, 256
   global, 40, 42, 44–45, 47
   history of, 50, 170, 191, 196, 206, 211, 251–52, 276
   invisible, 9, 12, 17
   limits of, 44, 45, 48
   material, 39, 120, 167, 173, 179, 215, 237
   model of, 169
   observable/unobservable, 39–44, 47, 156
   order of, 221
   origin of, 14, 16, 68, 73–76, 79, 81, 83, 173, 182–84, 186, 213, 218, 224, 230, 256, 260, 275
   physical, 9, 12–14, 16, 18, 38, 40–41, 44, 52, 55–56, 61, 65, 73, 179, 182–83, 185, 188, 196, 208, 212
   visible, 17, 148
Urban VIII, Pope, 59

Wassmann, Erich, 197
Watson, James, 37
William of Ockham, 27
Williams, Roger, 265
Wolff, Johann Christian, 28
world, 2, 12, 24, 26, 28, 31–32, 61–62, 75, 119, 124, 146, 182, 189, 196, 207–8, 236, 246, 257–58, 264, 268–69, 270
   consummation of, 19
   formation, 19
   material/immaterial, 5, 11, 18, 173, 206
   natural, *see* natural
   physical, 149
   real, 124, 149
   supernatural, 12
   visible/invisible, 10, 248
worldview, ix–x, 2, 33, 57, 59, 67, 169, 182, 184–85, 230, 235, 240, 269
   atheistic, 82
   biblical, 79
   Christian, 185, 234, 273
   materialistic, 74, 79, 234
   medieval, 28, 183, 270
   pagan, 185

Zahm, John A., 62, 196, 217, 243, 245
Zeno of Elea, 98, 113
Zycinski, Jozef, bishop, 189, 217, 251

www.ingramcontent.com/pod-product-compliance
Lightning Source LLC
Chambersburg PA
CBHW051630230426
43669CB00013B/2244